The SIOP® Model for Teaching Science to English Learners

Deborah J. Short
Center for Applied Linguistics, Washington, DC
Academic Language Research & Training, Arlington, VA

MaryEllen Vogt
California State University, Long Beach

Jana Echevarría
California State University, Long Beach

With contributions by
Amy Ditton
AIM Educational Consulting
Hope Austin-Phillips
Minneapolis Public Schools

Boston Columbus Indianapolis New York San Francisco Upper Saddle River
Amsterdam Cape Town Dubai London Madrid Milan Munich Paris Montreal Toronto
Delhi Mexico City Sao Paulo Sydney Hong Kong Seoul Singapore Taipei Tokyo

Vice President, Editor-in-Chief: *Aurora Martínez Ramos*
Series Editorial Assistant: *Meagan French*
Vice President, Director of Marketing: *Quinn Perkson*
Marketing Manager: *Danae April*
Production Editor: *Gregory Erb*
Editorial Production Service: *Nesbitt Graphics, Inc.*
Manufacturing Buyer: *Megan Cochran*
Electronic Composition: *Nesbitt Graphics, Inc.*
Interior Design: *Nesbitt Graphics, Inc.*
Photo Researcher: *Annie Pickert*
Cover Designer: *Linda Knowles*

For Professional Development resources visit www.pearsonpd.com.

Cataloging-in-Publication Data is on file at the Library of Congress.

22 17

Photo Credits: pp. 1, 114, 143: Bob Daemmrich Photography; p. 21:Bob Daemmrich/The Image Works; p. 52: Brand X/Jupiter Images/Getty RF; p. 79: Frank Siteman; p. 102: Photos To Go; p. 128: 67photo/Alamy; p. 157: Harmik Nazar/Hill Street Studios/Getty Images.
All photos in Appendix C appear courtesy of iStockphoto.com.

www.pearsonhighered.com

ISBN-10: 0-205-62759-5
ISBN-13: 978-0-205-62759-2

Dedication

This book is dedicated to all the teachers who are committed to teaching English learners the content and language of science using the SIOP® Model. May you discover new learnings for yourself within. This book is for YOU!

contents

preface

We have written this book in response to the many requests from teachers of science for specific application of the SIOP® Model to science. During our nearly fifteen years of working with the SIOP® Model, we have learned that both the subject you teach and the audience for whom you teach it is a major consideration. Showing a SIOP® social studies lesson plan to a chemistry teacher, along with the advice to "adapt it," has sometimes resulted in rolled eyes and under-the-breath comments like, "You've got to be kidding." A similar reaction has occurred when we have shown a physical science lesson video clip to elementary reading teachers and asked them to modify the techniques in their classes. Many teachers, whatever the subject area, prefer to see examples specific to their content and grade level.

So, this book is intended specifically for teachers of science content or language, including elementary classroom teachers, secondary science teachers, ESL specialists, and SIOP® coaches. If you teach in grades K–2, 3–5 (or 6), 6–8, or 9–12, you'll find information about teaching science written specifically for your grade-level cluster. If you are an elementary teacher, an ESL specialist, SIOP® coach, or teacher educator, you may teach or support multiple subjects and so you may want to check out our companion books for teaching mathematics, social studies, and English-language arts within the SIOP® Model.

We offer an important caveat. This book has been written for teachers who have familiarity with the SIOP® Model. Our expectation is that you have read one of the core texts: *Making Content Comprehensible for English Learners: The SIOP® Model* (Echevarría, Vogt, & Short, 2008); or either *Making Content Comprehensible for Elementary English Learners: The SIOP® Model* (Echevarría, Vogt, & Short, 2010a) or *Making Content Comprehensible for Secondary English Learners: The SIOP® Model* (Echevarría, Vogt, & Short, 2010b). If you have not read one of these books or had substantial and effective professional development in the SIOP® Model, we ask that you save this book for later. We want this book to be just what you've been looking for, a resource that will enable you to more effectively teach science to your English learners and other students. Therefore, the more familiar you are with the philosophy, terminology, concepts, and teaching techniques associated with the SIOP® Model, the better you will be able to use this book. If you would like a brief refresher on the SIOP® Model, please read Appendix A, SIOP® Protocol and Component Overview.

The SIOP® Model is the only empirically validated model of sheltered instruction at present. Sheltered instruction or SDAIE (Specially Designed Academic Instruction in English), in general, is a means for making content comprehensible for English learners (ELs) while they are developing English proficiency. The SIOP® Model distinctively calls for teachers to promote academic language development along with comprehensible content. SIOP® classrooms may include a mix of native-English speaking students and English learners, or just English learners. This depends on your school and district, the number of ELs you have in your school, and the availability of SIOP®-trained teachers. Whatever your context, what characterizes a sheltered SIOP® classroom is the systematic, consistent, and concurrent focus on teaching both academic content and academic language to English learners.

This book is intended to deepen your understanding of the SIOP® Model and provide more specific teaching ideas, lesson plans, and comprehensive unit plans for teaching science to English learners. Our overall goal is to help you master SIOP® lesson and unit planning, enabling you to incorporate the components and features of the SIOP® Model consistently in your classroom.

Organization and Purpose of This Book

We have specifically written this book for you, elementary and secondary science educators who have already studied the SIOP® Model, read *Making Content Comprehensible for English Learners: The SIOP® Model* (Echevarría, Vogt, & Short, 2008), and are ready to enhance your SIOP® lessons with specific techniques that target your content area. In the book, you will read about a wide variety of instructional activities, many of which are effective for any grade level or science course. The readability of the textbooks, the depth and complexity of the concepts, and the language demands of the assignments change, of course, as the grade levels rise, but nearly all of the meaningful activities showcased here will work well with all students, including ELs, in grades K–12, enabling them to explore, practice, and apply key concepts and academic language. Therefore, we encourage you to review the lessons and units from grade-level clusters other than the one in which you teach. Throughout all of the techniques in Chapters 2 and 3 and the units and lesson plans in Chapters 5–8, you will find many teaching ideas and activities that truly span all grade levels and content areas, so please don't overlook them.

Chapter 1: The Academic Language of Science

Chapter 1 focuses on the academic language that students need to be successful in school. While it is true that ELs benefit from similar vocabulary learning strategies as other students, they generally need more explicit support in vocabulary development (more techniques, for example, that use realia and demonstrations, highlight cognates, identify words with multiple meanings, and focus on idiomatic speech). They need direct instruction in other aspects of academic language as well, namely the academic reading, writing, and oral discourse skills that characterize a science class. These aspects may be broad-based uses of language, such as how to record observations, take notes from a lecture or reference material, justify orally or in writing a conclusion resulting from a science experiment, or construct an argument in a debate setting. Other aspects of academic language are more narrow and related to English grammar and usage, such as using transitions and cause-effect language properly, writing conditional sentences, and comprehending the passive voice. In this chapter, we also extract examples of how academic English is used in science, so you can be more aware of potential pitfalls for ELs as well as language teaching opportunities.

Chapters 2 and 3: Activities and Techniques for Planning SIOP® Science Lessons

Chapters 2 and 3 describe effective techniques that teachers may use in SIOP® science lessons and apply them to representative classrooms. Some of these techniques have been drawn from the *99 Ideas and Activities for Teaching with the SIOP® Model* (Vogt & Echevarría, 2007) and others were suggested or created by our contributors, SIOP®-trained science educators Amy Ditton (elementary grades) and Hope Austin-Phillips (secondary grades). Many are not new techniques, but we have specialized them for science topics and included scenarios that demonstrate their application at particular grade levels in subjects different from those in the later science units. As you read through the lesson applications in these chapters, pay close attention to how language is embedded into, for example, the scientific concepts of biomes and elements on the periodic table. You may be surprised how much language is involved in teaching scientific concepts once you focus on it.

Chapter 4: SIOP® Science Lesson Planning and Unit Design

Chapter 4 focuses on SIOP® science unit design and lesson planning. We discuss guidelines for planning SIOP® science lessons and offer a model with a unit planner to show how teachers can build a week-long science unit that not only covers the regular curriculum but also enables students to make progress with their language development in all four domains over the course of several days. We describe the lesson plan format used in this book and explain how to incorporate the techniques and activities from Chapters 2 and 3. Next we include sample lesson plans that more fully apply two or three of the techniques previously described. By reviewing these plans, you will see how a SIOP® science lesson can be successfully developed.

Chapters 5–8: Sample SIOP® Science Units and Lessons

In these chapters, four units are illustrated, one each for Grades K–2 (Chapter 5), 3–5 (Chapter 6), 6–8 (Chapter 7), and 9–12 (Chapter 8). Many science topics are taught in different grades across the United States so we felt the cluster approach would be most useful. Our contributors describe their planning process for each unit presented, discussing the objectives they highlighted and the standards they derived lessons from, the SIOP® techniques and activities they have chosen and placed across the five lessons, their selection of materials, and other goals they have.

You will find several "Think-alouds" and "Planning Points" throughout these lesson plans through which the writers convey their decision-making process, such as when to teach key vocabulary terms and how to elicit more academic language from the ELs when they are working in groups. For those of you unfamiliar with "Think-alouds," they are structured models of how successful readers, writers, and learners think about language and learning tasks (Baumann, Jones, & Seifer-Kessel, 1993; Oczkus, 2009). The "Planning Points" comments clarify and provide additional information, including planning tips. You will find lessons from Chapter 4 embedded in these units. Being familiar with the topic and one lesson already should allow you to envision the delivery of the unit more fully. The lesson plans also include handouts the teachers might use with students, such as specific graphic organizers and charts.

Chapter 9: Pulling It All Together

In Chapter 9, we conclude the book with some of our own thoughts, insights, and recommendations, as well as those from the content specialists and SIOP® experts who served as contributors to this book. We hope this chapter will help you pull it together as you continue your journey to effective SIOP® instruction.

Appendices

Several resources are available in the appendices. Appendix A provides an overview of the SIOP® Model. Appendix B offers a sample of the academic science words found in state standards. Appendix C contains blackline masters of student handouts for the lessons found in Chapters 5–8. Appendix D offers two versions of the SIOP® Model protocol that can be used for observing and coaching SIOP® lessons.

To further assist you in creating a successful SIOP® classroom, remember that several other resources are also available. These include (in addition to the core texts mentioned previously): *99 Ideas and Activities for Teaching English Learners with the SIOP® Model* (Vogt & Echevarría, 2008); *Implementing the SIOP® Model through Effective Professional Development and Coaching* (Echevarría, Short, & Vogt, 2008); and *The SIOP® Model for Administrators* (Short, Vogt, & Echevarría, 2008).

Acknowledgments

We acknowledge and appreciate the suggestions offered by the educators who reviewed this book. They include Margaret McGowan, Egg Harbor Township School District; Emily Morgan, Fairmount Junior High School, Boise; and Judith B. O'Loughlin, New Jersey City University.

We would also like to thank Jennifer Himmel and Cara Richards for their commitment to the SIOP® Model in science and their research efforts in that area and the staff of the Creighton (AZ) School District Language Acquisition Department for their support and mentoring of SIOP® Model instruction.

To our Allyn & Bacon team we express our gratitude for keeping us focused and on task and for its dedication to sharing information about the SIOP® Model. To Aurora Martínez, our incredible editor who never sleeps, we know that it has been through your understanding of the academic and language needs of the English learners in our schools that this series of books has been written on the SIOP® Model. Thank you, Aurora.

We have been most fortunate to have as our contributors to these content area books eight content specialists and SIOP® experts. Their insights, ideas, lesson plans, and unit plans across the grade level clusters clearly demonstrate their expertise not only in their content areas, but also in the SIOP® Model. With deep gratitude we acknowledge the significant contributions of our colleagues: Hope Austin-Phillips and Amy Ditton (Science); Aracelli Avila and Melissa Castillo (Mathematics); Robin Liten-Tejada and John Seidlitz (History-Social Studies); and Karlin LaPorta and Lisa Mitchener (English-Language Arts). Their deep understanding of the SIOP® Model shines through in their teaching techniques, lesson plans, and units, and we thank them for their belief in and commitment to English learners and the SIOP® Model.

Finally, we acknowledge and express great appreciation to our families and the families of our contributors. The SIOP® Model would not exist if it hadn't been for their support and encouragement over all these years.

djs
mev
je

about the authors

Deborah J. Short is a professional development consultant and a senior research associate at the Center for Applied Linguistics in Washington, DC. She co-developed the SIOP® Model for sheltered instruction and has directed national research studies on English language learners funded by the Carnegie Corporation, the Rockefeller Foundation, and the U.S. Dept. of Education. She recently chaired an expert panel on adolescent ELL literacy. As the director of Academic Language Research & Training, Dr. Short provides professional development on sheltered instruction and academic literacy around the U.S. and abroad. She has numerous publications, including the SIOP® book series and five ESL textbook series for National Geographic/Hampton-Brown. She has taught English as a second/foreign language in New York, California, Virginia, and the Democratic Republic of Congo.

MaryEllen Vogt is Distinguished Professor Emerita of Education at California State University, Long Beach. Dr. Vogt has been a classroom teacher, reading specialist, special education specialist, curriculum coordinator, and university teacher educator. She received her doctorate from the University of California, Berkeley, and is a co-author of fourteen books, including *Reading Specialists and Literacy Coaches in the* Real *World* (2nd ed., 2007) and the SIOP® book series. Her research interests include improving comprehension in the content areas, teacher change and development, and content literacy and language acquisition for English learners. Dr. Vogt was inducted into the California Reading Hall of Fame, received her university's Distinguished Faculty Teaching Award, and served as President of the International Reading Association in 2004–2005.

Jana Echevarría is a Professor Emerita at California State University, Long Beach. She has taught in elementary, middle and high schools in general education, special education, ESL and bilingual programs. She has lived in Taiwan, Spain and Mexico. Her UCLA doctorate earned her an award from the National Association for Bilingual Education's Outstanding Dissertations Competition. Her research and publications focus on effective instruction for English learners, including those with learning disabilities. Currently, she is Co-Principal Investigator with the Center for Research on the Educational Achievement and Teaching of English Language Learners (CREATE) funded by the U.S. Department of Education, Institute of Education Sciences (IES). In 2005, Dr. Echevarría was selected as Outstanding Professor at CSULB.

About the Science Contributors

Amy Ditton began her teaching career in Phoenix, Arizona, working in diverse classrooms with English learners. She learned the SIOP® Model as a middle school teacher and has gone through the process of collaborating with others to implement the model in her own classroom and K–8 school district. She now trains and works with teachers and administrators around the country to support schools in understanding second language acquisition and implementing the SIOP® Model.

Hope Austin-Phillips has been teaching middle school science in diverse classrooms for six years. She is passionate about teaching English learners and partially attributes this to learning how to implement the SIOP® Model. While teaching in Phoenix, Arizona, Hope also worked as a coach helping other teachers utilize the SIOP® Model. She is currently teaching in Minneapolis, Minnesota.

chapter 1

The Academic Language of Science

In most states, English learners (ELs) are struggling to meet academic targets in science set by the No Child Left Behind Act. One factor contributing to the difficulty ELs experience is that science is more than just doing experiments and observing natural phenomena; science education involves technical terms and associated concepts, oral or written instructions on how to conduct laboratory experiments, classroom academic language used in a teacher's explanation of a process, textbook reading, and report writing. Language plays a large and important role in learning science.

Consider the following situation. Your sister called earlier with some horrible news. Your three-year-old nephew has been diagnosed with Wilms' tumor. You have never heard of it and even your sister's explanation that it is a type of child's kidney disease, a cancer, doesn't ring any bells. Knowing you teach science, she is asking for your help in understanding the disease and the possible treatments.

You immediately begin with Internet searches. *Wikipedia* (Retrieved June 26, 2009 from http://en.wikipedia.org/wiki/Wilms'_tumor) gives you the following:

> Wilms' tumor or nephroblastoma is a tumor of the kidneys that typically occurs in children, rarely in adults. Pathologically, a triphasic nephroblastoma comprises three elements:
>
> blastema
>
> mesenchyme
>
> epithelium
>
> Wilms' tumor is a malignant tumor containing metanephric blastema, stromal and epithelial derivatives. Characteristic is the presence of abortive tubules and glomeruli surrounded by a spindled cell stroma. The stroma may include striated muscle, cartilage, bone, fat tissue, fibrous tissue. The tumor is compressing the normal kidney parenchyma.
>
> The mesenchymal component may include cells showing rhabdomyoid differentiation. The rhabdomyoid component may itself show features of malignancy (*rhabdomyosarcomatous Wilms*).

This doesn't help you understand the disease very much. You don't know many of the technical terms—*blastema, mesenchyme, metanephric, stromal, glomeruli*, and *rhabdomyoid*—although you do recognize that some sentences are formulated like definitions: *Wilms' tumor IS a malignant tumor containing. . . , The stroma MAY INCLUDE. . . .* It makes you remember a workshop you attended a few weeks ago in which the facilitator said that research reveals that we can only read independently if we know 90%–95% of the words.[1] In reading "The mesenchymal component may include cells showing rhabdomyoid differentiation," you know seven of the nine words, or 78%, but not the key ones that would convey meaning. What's *mesenchymal*? What's *rhabdomyoid*? This isn't going to help your sister and nephew determine a course of action without additional support.

So you read on and find some good news, "It is highly responsive to treatment, with about 90% of patients surviving at least five years." You continue reading eagerly, but the prognosis section doesn't help you understand what to do or how to help your nephew:

> Tumor-specific loss-of-heterozygosity (LOH) for chromosomes 1p and 16q identifies a subset of Wilms tumor patients who have a significantly increased risk of relapse and death. LOH for these chromosomal regions can now be used as an independent prognostic factor together with disease stage to target intensity of treatment to risk of treatment failure.[3][4] Genome-wide copy number and LOH status can be assessed with virtual karyotyping of tumor cells (fresh or paraffin-embedded).

This is frustrating. You are accustomed to reading science texts. You can figure out general academic words like *identifies, subset, increased, risk, regions,* and others, but you can't determine whether this paragraph is useful in your nephew's circumstances. Does he have an increased risk of relapse and death? What is loss-of-heterozygosity? If something is favorable, why would the text indicate it leads to increased risk of death?

Despite being well read and well educated, we have all had experiences where we became lost when listening to or reading about a new and unfamiliar topic. We're tripped

[1]Nagy & Scott, 2000.

up by the terminology, phrases, and concepts that are unique to the subject matter. When this happens, we most likely become frustrated and sometimes lose interest. But in this particular case, you don't want to give up because you want to help your nephew. Further, with your university education, you know how to access additional resources and how to reach out to knowledgeable experts for the information you and your sister will need.

However, every day, many English learners sit in classrooms where the topic, the related words, and concepts are totally unfamiliar to them. Other English learners may have familiarity with the topic, perhaps even some expertise, but because they don't know the English words, terminology, and phrases—that is, the content-specific academic language—they are also unable to understand what is being taught. Comprehension can be compromised as well when they don't understand cause-effect sentence structures or the usage of such prepositions and conjunctions as *except*, *unless*, *but*, *despite*, or *however*. Moreover, they have not yet mastered how to use language and content resources to help them understand.

What Is Academic Language?

Although definitions in the research literature differ somewhat, there is general agreement that academic language is both general- and content-specific. That is, many academic words are used across all content areas (such as *demonstrate, conflict, analyze, element, category*), whereas others pertain to specific subject areas (*photosynthesis, mitosis, density, hypothesize,* and *inertia* for science; *angle, ratio, dispersion,* and *calculate* for math). It is important to remember that academic language is more than specific content vocabulary words related to particular topics. Rather, academic language represents the entire range of language used in academic settings, including elementary and secondary schools. Consider the following definitions offered by several educational researchers:

- Academic language is "the language that is used by teachers and students for the purpose of acquiring new knowledge and skills . . . imparting new information, describing abstract ideas, and developing students' conceptual understandings" (Chamot & O'Malley, 1994, p. 40).
- Academic language refers to "word knowledge that makes it possible for students to engage with, produce, and talk about texts that are valued in school" (Flynt & Brozo, 2008, p. 500).
- "Academic English is the language of the classroom, of academic disciplines (science, history, literary analysis) of texts and literature, and of extended, reasoned discourse. It is more abstract and decontextualized than conversational English" (Gersten, Baker, Shanahan, Linan-Thompson, Collins, & Scarcella, 2007, p. 16).
- Academic English "refers to more abstract, complex, and challenging language that will eventually permit you to participate successfully in mainstream classroom instruction. Academic English involves such things as relating an event or a series of events to someone who was not present, being able to make comparisons between alternatives and justify a choice, knowing different forms, and inflections of words and their appropriate use, and possessing and using content-specific vocabulary and modes of expression in different academic disciplines such as mathematics and social studies" (Goldenberg, 2008, p. 9).

- "Academic language is the set of words, grammar, and organizational strategies used to describe complex ideas, higher-order thinking processes, and abstract concepts" (Zwiers, 2008, p. 20).

When you reflect on the examples for science and mathematics, you can see that academic language differs considerably from the social, conversational language that is used on the playground, at home, or at cocktail parties. Social or conversational language is generally more concrete than abstract, and it is usually supported by contextual clues, such as gestures, facial expressions, and body language (Cummins, 1979, 2000; Echevarria & Graves, 2007).

Some educators suggest that the distinction between conversational and academic language is somewhat arbitrary and that it is the *situation, community,* or *context* that is either predominantly social or academic (Aukerman, 2007; Bailey, 2007). For purposes of this book, we maintain that academic language is essential for success in school and that it is more challenging to learn than conversational English, especially for students who are acquiring English as a new language. Although knowing conversational language assists students in learning academic language, we must explicitly teach English learners (and other students, including native speakers) the "vocabulary, more complex sentence structures, and rhetorical forms not typically encountered in nonacademic settings" (Goldenberg, 2008, p. 13).

A focus on words, grammar, and oral and written discourse as applied in school settings is likely to increase student performance levels. Analyses of language used in assessments by Bailey and Butler (2007) found two types of academic language: content-specific language (e.g., technical terms such as *latitude* and *longitude*, and phrases such as "We hypothesize that . . .") and general, or common core, academic language (e.g., persuasive terms, comparative phrases) that is useful across curricular areas. Similarly, there are general academic tasks that one needs to know how to do in order to be academically proficient (e.g., create an outline, take notes) and more specific tasks (e.g., write a scientific laboratory report). They argue that teachers and curricula should pay attention to this full range of academic language and that the enhancement of ELs' academic language skills should enable them to perform better on assessments. This conclusion is bolstered by Snow, et al. (1991), who found that performance on highly decontextualized (i.e., school-like) tasks, such as providing a formal definition of words, predicted academic performance, whereas performance on highly contextualized tasks, such as face-to-face communication, did not.

How Is Academic Language Manifested in Classroom Discourse?

Our teachers come to class,
And they talk and they talk,
Til their faces are like peaches,
We don't;
We just sit like cornstalks.

(Cazden, 1976, p. 74)

These poignant words come from a Navajo child who describes a classroom as she sees it. Teachers like to talk. Just observe any classroom and you'll find that the teacher does the vast majority of the speaking. That might be expected because the teacher, after all, is the most expert science person in the classroom. However, for students to develop proficiency in language, interpret what they read and view, express themselves orally and in writing, participate during whole-group instruction and small-group interaction, and explain and defend their answers, they need opportunities to learn and use academic language.

Many of the visible manifestations of academic language use in the classroom come from the conversations between teacher and students, and on occasion among students. Most instructional patterns involve the teacher initiating a topic (I) usually by asking a question, a student responding (R), the teacher evaluating (E) the response or providing feedback (F), followed by another teacher-asked question (Cazden, 1986, 2001; Mehan, 1979; Watson & Young, 1986). A typical interaction between a teacher and her students during a science lesson is illustrated in the following example:

T: Who can name one of the three types of rocks we studied yesterday?

S: Igneous.

T: Right. Igneous rock comes from volcanoes. Who can tell us another type?

S: Sed-, sedi-, sedimentary.

T: That's right. Good. This rock type is a result of little bits of rocks and sand pressing together in layers over time.

And on it goes, often for a good portion of the lesson. Notice that the teacher asked questions that had a correct answer with no reasoning or higher level thinking required; in fact, the verb "name" in the teacher's question indicates she is asking for a factual recall. The teacher controlled the interchange, and the teacher evaluated student responses. Also note that the only person in the interchange to actually orally produce elaborated academic language (in this case, definitions of the types of rocks) was the teacher. The students didn't need to use more than one or two words in response to the teacher's questions in order to participate appropriately. But it is the students who need to practice using academic language, not the teacher! Further, only two students were involved; the others were quiet.

The Initiation-Response-Evaluation/Feedback (IRE/F) pattern is quite typical and it has been found to be one of the least effective interactional patterns for the classroom (Cazden, 1986, 2001; Mehan, 1979; Watson & Young, 1986). More similar to an interrogation than to a discussion, this type of teacher–student interaction stifles academic language development and does not encourage higher level thinking because most of the questions have a straightforward, known answer. Further, we have observed from kindergarten through high school that most students become conditioned to wait for someone else to answer. Often it is the teacher who ultimately answers his or her own question, if no students volunteer. Or the teacher elaborates, as in the third and fifth lines above.

In classrooms where the IRE/F pattern dominates, the teacher's feedback may inhibit learning when he or she changes students' responses by adding to or deleting from their statements or by completely changing a student's intent and meaning. Because the teacher is searching for a preconceived answer and often "fishes" until it is found, the cognitive work of the lesson is often carried out by the teacher rather than the students. In these classrooms, students are seldom given the opportunity to elaborate on their answers; rather, the teacher does the analyzing, synthesizing, generalizing, and evaluating.

Changing ineffective classroom discourse patterns by creating authentic opportunities for students to develop academic language is critically important because as one acquires language, new concepts are also developed. Think about the experience described previously regarding websites with information about Wilms' tumor. Each new vocabulary word you learned and understood (e.g., metanephric blastema, loss-of-heterozygosity) is attached to a concept that in turn expands your ability to think about and begin to understand your nephew's disease and consider courses of action. As your own system of word-meaning grows in complexity, you are more capable of understanding the associated concepts and generating the self-directed speech of verbal thinking, such as "If we can determine my nephew's loss of heterozygosity for certain chromosome areas, we'll have a sense of whether the treatment might work in the long run." Without an understanding of the words and the concepts they represent, you would be incapable of thinking about (self-directed speech) or discussing (talking with another) Wilms' tumor disease.

Academic English also involves reading and writing. As you most likely know, the National Reading Panel (National Institute of Child Health and Human Development, 2000) defined the major components of reading as phonics, phonemic awareness, fluency, vocabulary, and reading comprehension. Research suggests that high-quality instruction in these five components generally works for English learners as well, although additional focus on oral language development and background building are called for to enhance comprehension (August & Shanahan, 2006; Goldenberg, 2008).

Although English learners are able to attain well-taught word-level skills such as decoding, word recognition, and spelling that are equal to their English-speaking peers, the same is not typically the case with text-level skills such as reading comprehension and writing (Goldenberg, 2008). One reason for the disparity between word-level and text-level skills among English learners is oral English proficiency. Well-developed oral proficiency in English, which includes English vocabulary and syntactic knowledge plus listening comprehension skills, is associated with English reading and writing proficiency. Therefore, it is insufficient to teach English learners the components of reading alone; teachers also must incorporate extensive oral language development opportunities into literacy instruction. Further, English learners benefit from more opportunities to practice reading, check comprehension, and consolidate text knowledge through summarization. They also need instruction on the features of different text genres, especially those found in subject area classes—such as textbook chapters, online articles, laboratory directions, diagrams and other graphics, and primary source materials. Since reading is the foundation for learning in school, it is critical that teachers use research-based practices to provide English learners with high-quality instruction that will lead to the development of strong reading skills.

Academic writing is an area that is affected significantly by limited English proficiency. While oral skills can be developed as students engage in meaningful activities, skills in writing must be explicitly taught. The writing process, which involves planning, drafting, editing, and revising written work, allows students to express their ideas at their level of proficiency with teacher (or peer) guidance and explicit corrective feedback. However, for English learners, it is critical that a lot of meaningful discussion take place prior to asking students to write because such dialogue helps connect ideas in support of writing and provides students with the English words they will use. Writing is also facilitated by such things as teacher modeling, posting of writing samples, providing sentence frames, and even having students copy words or text until they gain more independent proficiency (Graham & Perin, 2007). This kind of constant exposure to words and sentence patterning allows ELs to become familiar with the conventions of how words and sentences are put together in the language (Garcia & Beltran, 2003).

English learners should be encouraged to write in English early, especially if they have skills in their native language, and should be provided frequent opportunities to express their ideas in writing. Errors in writing are to be expected and should be viewed as part of the natural process of language acquisition. Providing scaffolded writing tools, such as partially completed graphic organizers for pre-writing and sentence frames for organizing key points and supporting details will help ELs write in the content classroom.

What Is the Academic Language of Science?

A focus on teaching the language of science is not new, but it is still not widespread in practice, although *Science for English Language Learners* (Fathman & Crowther, 2006), *Teaching Science to English Language Learners* (Rosebery & Warren, 2008), and *Making Science Accessible for English Learners* (Carr, Sexton, & Lagunoff, 2007) are recent, useful resources. In the 1980s, Lemke began examining science discourse and its implications for students learning English as a new language. At that time, he explained,

> [E]ducators have begun to realize that the mastery of academic subjects is the mastery of their specialized patterns of language use, and that language is the dominant medium through which these subjects are taught and students' mastery of them tested. (Lemke, 1988, p. 81)

Lemke (1990) pointed out that "scientific language has a preference in its grammar for using the passive voice . . . people tend to disappear from science as actors or agents . . . [and there is] a grammatical preference for using abstract nouns derived from verbs" (p. 130). He criticized these stylistic conventions because they made science less accessible and less engaging to students. "(T)eachers tend to leave much of the semantics and grammar of scientific language completely implicit" (Lemke, 1990, p. 170). Criticism like Lemke's is one of the reasons we developed the SIOP® Model—to help make language explicit in content area classrooms.

Both Lemke (1990) and Gibbons (2003) have argued that teachers should talk directly with students about scientific discourse, introduce semantic relationships among scientific terms, and give students more practice in speaking about science. They have recommended using informal or everyday speech initially so students understand the information and concepts being taught, and teaching the ELs the necessary technical terms, grammatical expressions, and discourse patterns (such as use of argumentation afterwards). A recent, small-scale experimental study of a four-hour web-based lesson supports those recommendations. Brown and Ryoo (2008), who gave fifth graders a researcher-developed assessment of photosynthesis using scientific language, found that students who were taught with a method that explained scientific concepts in everyday language before introducing scientific terms performed better than students who were taught with a method that used the scientific terms from the outset.

FIGURE 1.1 *Hiebert's Challenges and Assets of Learning Science Vocabulary (Hiebert, 2008)*

Challenges	*Assets*
1. Dense	1. Clear delineation of vocabulary
2. Conceptually difficult	2. Build-up of ideas
3. Central to text	3. Concepts can be taught thematically
4. General academic vocabulary	4. Many clear Spanish cognates
5. Not much time for science instruction	5. Potential for high levels of engagement

Hiebert (2008) offers a useful description of the challenges and assets that exist in learning science vocabulary. As shown in Figure 1.1, the first four challenges for educators to keep in mind are related to academic language teaching. By the end of the elementary grades, science vocabulary is often found in dense text. This means that many scientific terms may be found in one given paragraph. The words are more often conceptually difficult to understand. In other words, a reader can't rely on known information to determine a meaning; rather, he or she must be taught new information that in turn helps explain the meaning. Science vocabulary is not easy to skip over. If you encounter an unknown word in a science text, it is usually central to the meaning of the sentence and often the entire paragraph. Skipping over and reading on is not as likely to help in comprehension as it might in other types of text, such as stories. Science, like many content areas, also uses a high number of more general academic words, such as *determine*, *explain*, and *predict*, which we discussed earlier in this chapter. The fifth challenge of too little time for science instruction has been true across the United States, with the increased emphasis on math and reading time, particularly in elementary schools. But we are seeing some adjustments now that science is part of the battery of assessments that Title I schools need to take under NCLB.

Hiebert (2008) has also identified aspects of science vocabulary that may facilitate learning for our English learners. Most of the science-specific terms, such as *velocity*, *xylem*, and *igneous*, have clear definitions that when learned aid in comprehension. Further, science concepts are built and reviewed over time. Unlike stories read in language arts, which can include very different sets of vocabulary for each story, science texts reinforce words within a grade and as students move up the grades. Repeated exposure and use of the words assists in learning. Science words can be taught thematically also. This type of instruction helps students remember and retrieve new terms. Many scientific terms and general academic words in English have close cognates in Latin-based languages; moreover, a number of the terms are high-frequency words in Spanish, but low-frequency words in English, which can benefit our ELs who are speakers of Spanish. The fifth asset is less directly related to vocabulary development but is important nonetheless for success in science classes. Science instruction can be highly engaging. If students are engaged in lessons, they are more likely to learn. That is one reason why student engagement is an important feature in the SIOP® Model.

Scott (1992) wrote "Language plays [roles] in science learning . . . science can be used to develop children's language, and . . . increased knowledge of language goes hand in hand with the development of scientific ideas" (p. ix). Researchers have found that students learn science better when they engage in literacy-related activities (Bredderman, 1983; Fellows, 1994; Holliday, Yore, & Alvermann, 1994; Rowe, 1996). In the classroom, then, science and language become interdependent. "These reciprocal skills give teachers and students a unique leverage: by merging science and language in the classroom, teachers can help students learn both more effectively" (Short & Thier, 2006, p. 206).

There are myriad terms that are used in academic settings. As mentioned previously, some of these are used commonly across the curricula and others are content specific. The metaphor of bricks and mortar may be helpful for you here if you think of some words representing bricks, such as science content-specific words (e.g., *electromagnet, meiosis, calcium chloride*). The mortar refers to general academic words (e.g., *determine, represent, attribute, approximate*) (Dutro & Moran, 2003). Understanding both types of terms is often the key to accessing content for English learners. For example, while most students need to have terms related to science explicitly taught, English learners also require that general academic words be included in vocabulary instruction. In addition, science often utilizes words with multiple meanings (i.e., polysemous words) for specific purposes and students may know one meaning, but not the other. Consider *table*, *mass*, *wave*, and *property*. ELs are likely to know of tables and chairs, a religious mass, a wave of the hand, and property their family owns, but may not know the scientific usage of these terms. So those terms need specific attention as well.

As you plan for lessons that teach and provide practice in both science-specific academic language and more general academic language, use your teacher's guides from your textbook to note the highlighted vocabulary, but consider other terms and phrases that may need to be taught. See Figure 1.2 by contributor Austin-Phillips for a breakdown of such terms and phrases. Also, you may use your state science content standards and English language proficiency standards to assist you in selecting the general academic language you need to teach and reinforce. Other resources include the "1,000 Most Frequent Words in Middle-Grades and High School Texts" and "Word Zones™ for 5586 Most Frequent Words," which were collected by Hiebert (2005) and may be found online at www.textproject.org. For those of you who are high school teachers, you might also want to take a look at the Coxhead Academic Word List (Coxhead, 2000). [Available at http://www.victoria.ac.nz/lals/staff/Averil-Coxhead/awl/ and http://simple.wiktionary.org/wiki/Wiktionary:Academic_word_list]

While studying science, therefore, students are exposed to new terms that they are unlikely to encounter in other subjects, general academic words that have use across the curriculum, and polysemous words for which they know a common meaning, but not the particular meaning used in the science context. Let's take a look at the various terms that are present in a few sample science standards. The words that are science specific are **bolded**, general academic words are underlined, and the polysemous words are in *italics*. Some words, you will see, are specific to science and also polysemous, so students may think they know what the words mean, but do not know the definition for the purpose intended by these standards.

In Kindergarten through Grade 2:

- Students know how to identify **resources** from Earth that are used in everyday life and understand that many resources can be **conserved**.
- Students know ***solids***, ***liquids***, and ***gases*** have different ***properties***.
- Students know many characteristics of an **organism** are **inherited** from the parents. Some characteristics are caused or influenced by the **environment**.

FIGURE 1.2 *Austin-Phillips' Science Language Chart*

Grade Level Band	*Technical Words*	*Process Words*	*Phrases/Sentence Starters*
K–2	natural, seasonal, living, non-living, environment, human, plant, moon, sun	similar, different, draw, label, question, cycle	• I notice . . . • I see . . . • I wonder . . . • Why does . . .? • How does . . .? • I want to know how/why . . . • Some characteristics of ____ are . . . • Some similarities are . . . • Some differences are . . .
3–5	refract, reflect, rotation, revolution, solid, liquid, gas, motion	characteristics, relationship, distinguish, observe, procedure, impact, argument	• I noticed . . . • These are similar because . . . • These are different because . . . • The data shows . . . • _________ is impacted by ______ because . . .
Middle School	mass, volume, particles, matter, mixture, pure substance, phenomena, ecosystem, weathering, erosion, deposition	influence, technology, logical, collaborate, diagram, infer	• One observation I made during the experiment was . . . • I observed . . . From my observations I can infer . . . • My hypothesis/prediction is supported/rejected because . . . • In conclusion . . . • We found that . . . • Our research question was . . . • My opinion is . . .
High School	cell theory, atomic theory, plate tectonic theory, isotopes, agitation, catalyst, thermodynamics, convection	skeptical, system, analysis, transfer, reliable, investigate	• According to the data . . . • The hypothesis was supported/rejected because . . . • Current research demonstrates . . . • In conclusion . . . • If_______, then . . . • The data indicates . . . • Possibilities for further research include . . . • Sources of error include . . .

In Grades 3–5:

- Students will collect **data** in an investigation and analyze those data to develop a logical conclusion.
- Students know plants are the *primary source* of ***matter*** and ***energy*** entering most **food chains**.
- Students know **electrically charged** objects attract or repel each other.

In Grades 6–8:

- Students will *plan* and conduct investigations in which ***independent*** and ***dependent variables***, ***constants***, ***controls***, and repeated *trials* are identified.

- Students will investigate and understand how **organisms** adapt to **biotic** and **abiotic** *factors* in a **biome**.
- Students will investigate and understand various *models* of **atomic structure** including **Bohr** and ***Cloud*** (**quantum**) models.

In Grades 9–12:

- Students know how to determine the **molar mass** of a **molecule** from its **chemical formula** and a *table* of **atomic masses** and how to convert the ***mass*** of a **molecular** substance to ***moles***, number of **particles**, or volume of ***gas*** at standard temperature and *pressure*.
- Students will be able to define **probability** and describe how it helps explain the results of **genetic crosses**.
- Students know how to identify **transverse** and **longitudinal** ***waves*** in **mechanical media**, such as *springs* and ropes, and on the Earth (**seismic waves**).

As you can see, many of the underlined words may be used in other content areas as well, but students need to be explicitly taught their specialized meaning in a particular science course. For students who speak a Latin-based language such as Spanish, cognates will help in teaching a number of words. For example, *predict* in English is *predecir* in Spanish; *justify* in English is *justificar* in Spanish; *investigation* in English is *investigación* in Spanish. Science-specific words should be explicitly taught as part of each science lesson.

You should also be aware that the national standards for English language proficiency (ELP) clearly state that students need to learn about science language (TESOL, 2006). They are similar to the WIDA (World-class Instructional Design and Assessment) standards that have been adopted by 22 states. The science language standard is:

> **English language learners communicate information, ideas, and concepts necessary for academic success in the content area of science.**

Model performance indicators are provided at five proficiency levels across grade-level clusters (PreK–K, 1–2, 3–5, 6–8, and 9–12) for the four domains—speaking, reading, writing, and listening. Gottlieb and Lederman (2006) suggest ways that national and state science standards can be integrated with these national ELP standards for science language and explain how the model performance indicators for the standards adjust the language load on students according to their proficiency levels, yet teach the curricular content.

In Appendix B you will find a listing of academic science vocabulary words found in the national and state science standards. Your state's standards and domains may differ somewhat, but we hope this extensive list will assist you in your lesson and unit planning, and in the writing of your content and language objectives.

Why Do English Learners Have Difficulty with Academic Language?

Developing academic language has proven to be quite challenging for English learners. In fact, in a study that followed EL students' academic progress in U.S. schools, researchers found that the ELs actually regressed over time (Suarez-Orozco, Suarez-Orozco & Todorova,

2008). There are a multitude of influences that affect overall student learning, and academic language learning in particular. Some factors, such as poverty and transiency, are outside of the school's sphere of influence, but some factors are in our control, namely what happens instructionally for these students that facilitates or impedes their learning.

Many classrooms are devoid of the kinds of supports that assist students in their quest to learn new material in a new language. Since proficiency in English is the best predictor of academic success, it seems reasonable that teachers of English learners should spend a significant amount of time teaching the vocabulary required to understand the lesson's topic. However, in a study that observed 23 ethnically diverse classrooms, researchers found that in the core academic subject areas only 1.4% of instructional time was spent developing vocabulary knowledge (Scott, Jamison-Noel, & Asselin, 2003).

The lack of opportunity to develop oral language skills hinders students' progress in all subject areas. Passive learning—sitting quietly while listening to a teacher talk—does not encourage engagement. In order to acquire academic language, students need lessons that are meaningful and engaging and that provide ample opportunity to practice using language orally. Successful group work requires intentional planning and teaching students how to work with others effectively; teacher expectations need to be made clear. Grouping students in teams for discussion, using partners for specific tasks, and other planned configurations increase student engagement and oral language development.

Another related reason that ELs struggle is lack of access to the language and the subject matter. Think about a situation in which you hear another language spoken. It could be the salon where you get a hair cut or your favorite ethnic restaurant. Just because you regularly hear another language, are you learning it? Typically not. Likewise, many English learners sit in class and hear what amounts to "English noise." It doesn't make sense to them and thus, they are not learning other academic language or the content being taught. Without the kinds of practices that are promoted by the SIOP® Model, much of what happens during the school day is lost on English learners.

We must also consider the types of classroom cultures students have experienced in the past. As Lemke (1990) noted, competence in content classes requires more than mastery of the subject matter topics; it requires an understanding of and facility with the genres and conventions for spoken and written interaction and the skills to participate in class activities. Currently, many science classes incorporate inquiry lessons that are designed to engage students in discovering scientific principles and conducting science experiments in a manner similar to the methods scientists use. However, some ELs who are recent immigrants may never have experienced an inquiry lesson. They may never have had the opportunity to conduct an experiment by manipulating scientific equipment and materials. They may have learned science through rote memorization of teacher lectures or textbook chapters. Therefore, teachers will need to introduce these ELs to a new classroom culture in which students are expected to participate orally, work in cooperative groups, solve problems, conduct experiments, generate hypotheses, express opinions, and so forth. Because the communication patterns in class may be very different from those in the students' native culture, teachers need to engage in culturally responsive teaching (Bartolomé, 1998), being sensitive to and building upon culturally different ways of learning, behaving, and using language. Working together, respectfully, the students and teacher can create a classroom culture in which they will all feel comfortable and learning can advance.

Finally, some teachers have low expectations for EL students (Lee, 2005). They are not motivated to get to know the students, their cultures, or their families. Poor performance is not only accepted, but expected. Rather than adjusting instruction so that it is meaningful to these students, teachers attribute lack of achievement to students' cultural background, limited English proficiency, and, sadly, ability. This attitude is unacceptable and staff who hold this view need to be re-educated in appropriate ways to teach these students and to learn that all students can reach high standards, although the pathways by which they attain them may vary.

How Can We Effectively Teach Academic Language with the SIOP® Model?

In a recent synthesis of existing research on teaching English language and literacy to ELs in the elementary grades, the authors make five recommendations, one of which is to "Ensure that the development of formal or academic English is a key instructional goal for English learners, beginning in the primary grades" (Gersten, et al., 2007, pp. 26–27). Although few empirical studies have been conducted on the effects of academic language instruction, the central theme of the panel of researchers conducting the synthesis was the importance of intensive, interactive language practice that focuses on developing academic language. This recommendation was made based upon considerable expert opinion, with the caveat that additional research is still needed. Additional reports offer similar conclusions (Deussen, Autio, Miller, Lockwood, & Stewart, 2008; Goldenberg, 2008; Short & Fitzsimmons, 2007).

Because you are already familiar with the SIOP® Model, you know that effective instruction for English learners includes focused attention on and systematic implementation of the SIOP® Model's eight components and thirty features. The SIOP® Model has a dual purpose: to systematically and consistently teach both content and language in every lesson. Content and language objectives not only help focus the teacher throughout a lesson, but also (perhaps even more importantly) focus students on what they need to know and be able to do during and after each lesson as related to *both* content knowledge and language development. Therefore, you should use the SIOP® protocol to guide lesson design when selecting activities and approaches for teaching academic language in your science courses. (See Echevarria & Colburn, 2006, for a discussion of designing inquiry-based SIOP® science lessons.)

Academic Vocabulary

Within the SIOP® Model, we refer to academic vocabulary as having three elements (Echevarria, Vogt, & Short, 2008, p. 59). These include:

1. *Content Words*: These are key vocabulary words, technical terms, and concepts associated with a particular topic. Key vocabulary, such as *solubility, covalent bond, ecosystem, mitochondria, Punnett square,* and *velocity,* typically come from science texts as well as from other components of the curriculum. Obviously, you will need to introduce and teach key content vocabulary when teaching about plants and animals, Physical Science, Earth Science, Biology, Chemistry, and Physics.
2. *Process/Function Words*: These are the words and phrases that have to do with functional language use, such as *how to make a hypothesis, provide evidence for a claim,*

state a conclusion, explain the effect, "state in your own words," summarize, ask a question, interpret, and so forth. They are general academic terms. Tasks that students are to accomplish during a lesson also fit into this category, and for English learners, their meanings may need to be taught explicitly. Examples include *list, explain, paraphrase, identify, create, monitor progress, define, share with a partner*, and so forth.

3. *Words and Word Parts That Teach English Structure*: These are words and word parts that enable students to learn new vocabulary, primarily based on English morphology. While instruction in this category generally falls under the responsibility of English-language arts teachers, we also encourage teachers of other content areas to be aware of the academic language of their own disciplines. The English-language arts (ELA) or English as a second language (ESL) teacher may teach the formation of the past tense (such as adding an *-ed* to regular verbs), yet you might reinforce past tense by pointing out that when we talk about scientific discoveries that happened in prior centuries, we use the past tense of English verbs, much like a history teacher might draw attention to past tense forms when discussing and reading about historical events. Similarly, when you give students written directions for lab experiments, you might point out that the steps tend to start with a verb, rather than a noun as found in basic sentences.

   ELA teachers will likely teach morphology (base words, roots, prefixes, suffixes), but you may teach many words with these word parts as key vocabulary (such as *in*vestiga*tion* or *bio*degr*ad*able). Science lends itself especially well to activities with roots and affixes because so many scientific terms utilize these word parts. Think about the root *derm*, for instance. If we teach students it means "skin," it might help them figure out *epidermis*, *dermatology*, *pachyderm*, and *hypodermic*, especially if we teach the suffix *-ology* and the prefix *hypo-*, too.

   For a usable and informative list of English word roots that provide the clue to more than 100,000 English words, refer to pages 60–61 of *Making Content Comprehensible for English Learners: The SIOP® Model* (Echevarria, Vogt, & Short, 2008). This is a must-have list for both elementary and secondary teachers in ALL curricular areas.

In sum, picture a stool with three legs. If one of the legs is broken, the stool will not function properly; it will not support a person who sits on it. From our experience, an English learner must have instruction in and practice with all three "legs" of academic vocabulary (content vocabulary, process/function words, and words/word parts that teach English structure) if they are going to develop the academic language they need to be successful students.

Zwiers (2008, p. 41) notes that "academic language doesn't grow on trees." Rather, explicit vocabulary instruction through a variety of approaches and activities provides English learners with multiple chances to learn, practice, and apply academic language (Stahl & Nagy, 2006). This requires teachers to provide comprehensible input (Krashen, 1985), as well as structured opportunities for students to produce academic language in their content classes. These enable English learners to negotiate meaning through confirming and disconfirming their understanding while they work and interact with others.

In addition to explicit vocabulary instruction, we need to provide a variety of scaffolds, including ones that provide context. Writing a list of science terms or pointing out terms that are in bold print in the textbook only helps if students know what they mean. To create a context for learning new words, teachers should preteach the terms, explain them in ways that students can understand and relate to, and then show how the terms are used in the textbook or classroom discourse. Scaffolding involves providing enough support to students so that the learners gradually are able to be successful independently.

Another way of scaffolding academic English is by having word walls or posters displayed that show key terms with visuals, definitions and/or sentences that use the term in context. In Chapter 2, one technique describes signal word posters that help students focus on words related to comparison and contrast, or cause and effect, or other relationships. Certainly, older learners can work in groups to create these posters with mnemonics, including cartoons or other illustrations. These aids reduce the cognitive load for English learners so that they can focus on scientific theory and processes without having to remember their associated linguistic terms. As students refer to and use these posted academic language words and phrases, the terms will become internalized and will later be used independently by students.

If English learners have opportunities to read, write, and orally produce words during science lessons and in their history, math, and/or English classes, the words are reinforced. And, if this reinforcement occurs throughout each and every school day, one can assume that English learners' mastery of English will be accelerated, much like repeated practice with any new learning.

Oral Discourse

Researchers who have investigated the relationship between language and learning suggest that there should be more balance in student talk and teacher talk to promote meaningful language learning opportunities for English learners (Cazden, 2001; Echevarria, 1995; Saunders & Goldenberg, 1992; Tharp & Gallimore, 1988; Walqui, 2006). In order to achieve a better balance, teachers need to carefully analyze their own classroom interaction patterns, the way they formulate questions, how they provide students with academic feedback, and the opportunities they provide for students to engage in meaningful talk.

Not surprisingly, teacher questioning usually drives the type and quality of classroom discussions. The IRE/F pattern discussed previously is characterized by questions to which the teacher already knows the answer and results in the teacher unintentionally expecting students to "guess what I'm thinking" (Echevarria & Silver, 1995). In fact, researchers have found explicit, "right there" questions are used about 50% of the time in classrooms (Zwiers, 2008) and science "discussions" can devolve into a series of factual exchanges.

In contrast, open-ended questions that do not have quick "right" or "wrong" answers promote greater levels of thinking and expression. During science lessons, there should be more of an emphasis on promoting classroom discourse by students questioning one another, separating fact from opinion, reasoning rather than memorizing procedures or guessing outcomes, making connections or generalizations, conducting experiments and communicating observations, drawing conclusions. For example, questions such as "Compare a cactus with a rose bush. Which is more suitable for a garden in your backyard and why?" and "Explain why Newton's first law of motion is important to the automobile industry" not only engender higher level thinking about scientific phenomena but also provide an opportunity for students to grapple with ideas and express themselves using academic English.

The Interaction component in SIOP® Model promotes more student engagement in classroom discourse. The features of the Interaction component, which should be familiar to you, include:

- Frequent opportunities for interaction with and discussion between teachers and students and among students, which encourage elaborated responses about lesson concepts
- Grouping configurations that support language and content objectives of the lesson
- Sufficient wait time for student responses that is consistently provided
- Ample opportunities for students to clarify key concepts in L1 (native language) as needed

These features promote balanced turn-taking between teachers and students, and among students, providing multiple opportunities for students to use academic English. Notice how each feature of Interaction encourages student talk. This is in considerable contrast to the discourse patterns typically found in both elementary and secondary classrooms.

Something as simple as having students turn to a partner and discuss an answer to a question first, before reporting out to the whole class, is an effective conversational technique, especially when the teacher circulates to monitor student responses. Speaking to a peer may be less threatening and also gets every student actively involved. Also, rather than responding to student answers with "Very good!" teachers who value conversation and discussion encourage elaborated responses with prompts such as "Can you tell us more about that?" or "What made you think of that?" or "Did anyone else have that idea?" or "Please explain how you figured that out."

Zwiers (2008, pp. 62–63) has classified comments teachers can make to enrich classroom talk; by using comments like these, a greater balance between student talk and teacher talk is achieved. Further, classroom interactions are less likely to result in an IRE/F pattern. Try using some of the comments below and see what happens to the interaction pattern in your own classroom!

To Prompt More Thinking

- You're on to something important. Keep going.
- You're on the right track. Tell us more.
- There is no right answer, so what would be your best answer?
- Can you connect that to something else you learned/saw/experienced?

To Fortify or Justify a Response

- That's a probable answer . . . How did you get to that answer?
- What evidence do you have to support that claim?
- What is your opinion/impression of . . . Why?

To Report on an Investigation

- Tell us more about what you noticed.
- Describe your result.
- How is your hypothesis the same or different?

- What do you think caused that to happen?
- How else might you study the problem?
- Can you generalize this to another situation? How?

To See Other Points of View

- So you didn't get the result you expected. What do you think about that?
- If you were in that person's shoes, what would you have done?
- Would you have done it like that? Why or why not?

To Consider Consequences

- Should she have . . . ?
- What if he had not done that?
- Some people think that . . . is wrong/right. What do you think? Why?
- How can we apply this to real life?

A conversational approach is particularly well-suited to English learners who, after only a few years in school, often find themselves significantly behind their peers in most academic areas, especially when instruction is in English, usually because of low reading levels in English, weaker vocabulary knowledge, and underdeveloped oral language skills. Students benefit from a conversational approach in many ways because conversation provides:

- A context for learning in which language is expressed naturally through meaningful discussion
- Practice using oral language, which is a foundation for literacy skill development
- A means for students to express their thinking, and to clarify and fine-tune their ideas
- Time to process information and hear what others are thinking about
- An opportunity for teachers to model academic language, use content vocabulary appropriately, and, through think-alouds, model thinking processes and learning strategies
- Opportunities for students to participate as equal contributors to the discussion, which provides them with repetition of both linguistic terms and thinking processes, and results in their eventual acquisition and internalization for future use

A rich discussion, or conversational approach, has advantages for teachers as well, including the following:

- Through discussion, a teacher can more naturally activate students' background knowledge and assess their prior learning.
- When working in small groups with each student participating in a discussion, teachers are better able to gauge student understanding of the lesson's concepts, tasks, and terminology, as well as discern areas of weakness.
- When teachers and students interact together, a supportive environment is fostered, which builds teacher–student rapport.

When contemplating the advantages of a more conversational approach to teaching, think about your own learning. In nearly all cases it takes multiple exposures to new terms, concepts, and information before you can use them independently. If you talk with others about the concepts and information you are learning, you're more likely to remember them. English learners require even more repetition and redundancy to improve their language skills. When they have repeated opportunities to improve their oral language proficiency, ELs are more likely to use English, and more frequent use results in increased proficiency (Saunders & Goldenberg, in press). With improved proficiency, ELs are more adept at participating in class discussions. Discussion and interaction push learners to think quickly, respond, construct sentences, put their thoughts into words, and ask for clarification through classroom dialogue. Discussion also allows students to see how other people think and use language to describe their thinking (Zwiers, 2008).

Productive discussion can take place in whole class settings, but it is more likely that small groups will facilitate the kind of high-quality interaction that benefits English learners. Working to express ideas and answers to questions in a new language can be intimidating for students of all ages. Small group work allows them to try out their ideas in a low-stress setting and to gauge how similar their ideas are to those of their peers. Working with partners, triads, or in a small group also provides a chance to process and articulate new information with less pressure than a whole class setting may create.

Earlier in this chapter, you read an interaction between a teacher and her students in which the IRE/F pattern prevailed. In contrast, read the following interaction from an eighth grade Physical Science class that was part of a SIOP® research study,[2] and reflect on the differences in the two classroom interaction patterns:

MS. ARMSTRONG: We saw the video clip and have been discussing the differences between chemical and physical change. Are you ready for your Round Robin activity?

STUDENTS: Yes. Yup. Uh hmm.

MS. ARMSTRONG: Okay, supply keepers come get your team's folders. This is Round Robin Classifying. Who remembers what to do?

ALYSHA: We number our papers 1–6. We get an index card with one of those numbers and answer the question. Then we pass the card to the next student and get a new card passed to us.

MS. ARMSTRONG: Do you agree or disagree with Alysha?

JORGE: I agree. It's like what we did last week with the ionic and covalent bonds.

MS. ARMSTRONG: Can you add any more to the directions?

JORGE: Well, we have only 1 minute to answer each one.

MS. ARMSTRONG: Okay. I'll tell you when to switch. Tick tock like a clock. You pass the cards clockwise.

[Students work in their teams answering the questions. Teacher calls out "Switch" after one minute passes and continues to do so until each student has a chance to answer all six questions.]

[2]All names are pseudonyms.

MS. ARMSTRONG: Now have a Roundtable discussion for five minutes. At your tables, what should I hear?

STUDENTS: Accountable talk.

MS. ARMSTRONG: Right. I want you thinking like a scientist.

[Green Team]

SONIA: I think the marshmallow is a chemical change.

TYRONE: Can you tell us why?

SONIA: Well, it changed temperature over the fire and got a hard crust on it.

ANDRES: I agree, it became a new substance.

TYRONE: Do we all agree? [Students nod.] Okay what about the water vapor?

ANDRES: The water is now a gas. That's a new substance too. [Teacher moves to this group and listens.]

DOLORES: Wait, I don't think that's right.

MS. ARMSTRONG: What's your evidence, Dolores?

DOLORES: Isn't water vapor still water? It's a gas but it's water too.

TYRONE: Can we think of a way to decide?

STEFANIE: When I cook beans, I put them in water in a pot. The water gets hot with bubbles. Water vapor goes up. But if I put the top on, drops of water are inside the top.

DOLORES: I see that too when I cook.

MS. ARMSTRONG: You're thinking outside the box.

ANDRES: So I'm wrong?

TYRONE: Yeah man, but it's okay. This isn't easy.

This class included ELs at all proficiency levels as well as native English speakers. Although all of your students may not sound exactly like these students, we know it is possible for them to participate in robust conversations as these students did. Much of the participation is internalized because the teacher has spent time with the class teaching accountable academic talk, has taught them routines for listening and responding to others, has encouraged them to respect one another's opinions, and has modeled how to justify an argument.

Note how the teacher facilitates this discussion with very few words: just some directions and then comprehension probes and careful listening. Teaching students to share conversational control and stepping back, trusting them to get the job done, takes some risk-taking on the part of the teacher and practice on the part of students, who may be used to just answering questions with monosyllabic responses. Simply telling students to "have a discussion among yourselves" will be less successful. We need to teach students how to engage in meaningful conversation and discussion and provide the support they need to do it well. Rather than sitting as "quiet cornstalks," students, including English learners, can learn to express themselves, support their viewpoints, advocate their positions, and defend their positions. When this occurs, we establish a classroom environment in which conversational control is shared among teachers and students alike.

Concluding Thoughts

Proficiency in English is the best predictor of academic success, and understanding academic language is an important part of overall English proficiency. In this chapter we have discussed what academic language is, why it is important, and how it can be developed in science classes and across the curriculum. In all content areas, teachers need to plan to explicitly teach both content area terms and general academic terms so that English learners can fully participate in lessons; acquire knowledge about concepts, theories, and processes of science; meet science standards; and increase their academic language proficiency.

For our students to achieve academically in science, they need to have practice with language skills that allows them to back up claims with evidence, be more detailed in their observations, use persuasive language compellingly in arguments, and compare events or points of view. When you teach students how to participate in classroom conversations and structured discussions, and how to read and write and think like a scientist, you not only improve their English skills but also prepare them for the academic language skills used in school and in professional settings. Teachers need to ensure that students internalize scientific habits of mind, such as using evidence to separate opinion from fact. If students are to become adults capable of making informed choices and taking effective action in the twenty-first century, then they must absorb those habits into their regular patterns of thought so the habits endure long after the students have graduated. Once they are scientifically literate, students will possess a set of skills that merges the knowledge of science concepts, facts, and processes with the ability to use language to articulate, converse about, and debate those ideas.

In the lesson plans and units that appear in Chapters 4–8, you will see a variety of instructional techniques and activities for teaching, practicing, and using academic language in science classrooms. As you read the lesson plans, reflect on why particular activities were selected for the respective content and language objectives. Additional resources for selecting effective activities that develop academic language and content knowledge include: Buehl's *Classroom Strategies for Interactive Learning* (2001); Vogt and Echevarria's *99 Ideas and Activities for Teaching English Learners with the SIOP® Model* (2008); Reiss's *102 Content Strategies for English Language Learners* (2008), and Marzano and Pickering's *Building Academic Vocabulary: Teacher's Manual* (2005). Secondary teachers will also find the following books, among many others, to be helpful: Zwiers's *Building Academic Language: Essential Practices for Content Classrooms (Grades 5–12)* (2008) and *Developing Academic Thinking Skills in Grades 6–12: A Handbook of Multiple Intelligence Activities* (2004); Fisher and Frey's *Word Wise and Content Rich: Five Essential Steps to Teaching Academic Vocabulary* (2008), and Cloud, Genesee and Hamayan's *Literacy Instruction for English Language Learners* (2009).

chapter 2

Activities and Techniques for SIOP® Science Lessons: Lesson Preparation, Building Background, Comprehensible Input, and Strategies

By Amy Ditton, Hope Austin-Phillips, and Deborah Short

Mr. Etherton's Vignette

Douglas Etherton is a fifth grade teacher in Phoenix, Arizona. His class is composed of twenty-three English learners, four bilingual/biliterate students, and four monolingual English speakers. His school is beginning the process of implementing the SIOP® Model, and Mr. Etherton will have support during the implementation with dedicated time to work with a site SIOP® coach and other teachers. Mr. Etherton teaches language arts,

math, science, and social studies. In the past, he did not spend as much time teaching science as reading and math, but this year, there is pressure from the administration to do more science.

In a conversation with his SIOP® coach, Mr. Etherton discussed his frustration with the science experiments that he had tried to incorporate in lessons. He shared the following with his coach, "The students aren't doing what I tell them to do during the experiments. They won't follow instructions and are doing the experiments incorrectly. They mix the wrong things together or they put things in the wrong place. They never record their observations in a way that makes any sense. When I give them a worksheet, the information they write down is also wrong or incomplete. All that seems to be happening is that they mix up the experiment, get the wrong results, and take notes that don't make sense. With all of the confusion, they really aren't understanding the science content."

When the coach, Mrs. Graham, asked him how he introduced the experiments, Mr. Etherton responded, "I have them read the directions, and I even have them repeat the directions back to me. When I ask them if they understand, they all say 'yes' and nod their heads. I give them time to ask questions, but they never have any before we get started. Once they do get started, it's chaos. They are fifth graders, they should be old enough to do this!"

After hearing about Mr. Etherton's challenges, Mrs. Graham asked what he was thinking of doing differently and he answered, "I guess they just can't handle doing experiments now. I really wanted to take a hands-on approach with science, but I'm not going to be able to. We'll have to go back to the textbook. Since they can't handle doing experiments or working in groups, they will have to read the chapter silently and answer the questions at the end. If they can do that without disruption, then I will also do some demonstrations for them."

Mrs. Graham empathized with Mr. Etherton's frustration, but encouraged him to keep trying experiments for a bit longer. She offered to conduct a mini-lesson with the students and also shared ideas for preparing students to work in groups, for modeling the science experiment steps in advance, and for providing them with graphic organizers that they could learn to record notes on.

Introduction

We know that simply adding experiments to a science lesson won't enable the students to do the work effectively, record their observations accurately, or draw the appropriate scientific conclusions. We also know that Mr. Etherton's notion of lecturing, reading, and answering questions from the textbook will not be successful either. Science can be such a motivating and engaging subject for students of all ages and all language backgrounds that Mr. Etherton should not give up! With the SIOP® coach's help, he will be able to incorporate new techniques and activities cohesively into meaningful lessons and scaffold how to do experiments, how to interpret the results, and how to use academic scientific language.

In this chapter and the following one, we present a variety of proven techniques and activities that teachers can use in their science lessons so that students will actively take part in the learning process, access the content, and develop proficiency in academic

English. You will see that after the steps for each technique are explained, the technique is applied in a classroom scenario that shows how it can be incorporated in a SIOP® science lesson. Knowing the challenges that accompany working with diverse groups of students, it was also important for us to consider not only that our students are second language learners but also that they have different language levels and academic needs. So, with the scenarios, we have also identified how the various techniques that are used to support the development of content and language can be modified for students at different proficiency levels or grade levels.

As you read and reflect on planning lessons yourselves, remember that the activities alone are not the lesson; not even doing an experiment should be an entire lesson. The focus should always be on the content and language objectives that the activities are meant to support. Select appropriate activities and techniques that allow students to practice new learning or reinforce new concepts, language structures, and processes.

Science Techniques and Activities

The activities and techniques in this chapter are organized around the first four SIOP® Model components: lesson preparation, building background, comprehensible input, and strategies. Chapter 3 presents techniques and activities for the final four components. We recognize that a number of the techniques are suitable for additional components, and have indicated those alternatives. The techniques may be used across various grade levels, K–12; they are not restricted to the grade level we discuss in the application scenarios or show in the sample lessons. Each technique's description identifies the optimal grades for use.

The following information is included for each technique in this chapter and Chapter 3.

Name of Activity or Technique – In cases where we have drawn from the techniques in the *99 Ideas* book, we indicate that here. In other cases, our contributors have included techniques and activities that they have named or ones that are familiar to many ESL educators.

SIOP® Component – Here, we identify the SIOP® component that this technique or activity addresses in the application we present. We recognize that several of these techniques may be used with different components as well.

Grade Level – Often, we provide a range of grade levels suitable for the technique or activity. While the classroom application is grade-specific, teachers will likely be able to modify the technique to a different grade within the range suggested.

Grouping Configurations – We explain what type of student configuration is most effective for this technique or activity. As you know, the way students are grouped is an important component of the SIOP® Model. We recommend deliberate, thoughtful groupings of students that match a lesson's objectives. Teachers may group students in pairs to promote more conversation and to lower anxiety levels. They may create small groups of students with different abilities to give more advanced students a chance to teach and less proficient students an opportunity to have peer role models. Teachers may at times group students by first language so the ELs can process the new information in a language they are more comfortable with before completing a task using English. There will also be times when a teacher wants to present new information to the whole group, but wants to ensure it is comprehensible and the students stay focused.

Approximate Time Involved – This information gives the teacher a sense of how long the activity may take in a lesson. It is useful for considering the pacing while preparing a lesson plan. It does not include the time needed to prepare for the activity.

Materials – Materials needed for the activity are listed here.

Description/Procedure – We describe the technique or activity here, including its general purpose and the steps of the procedure. Some suggestions for topics pertinent to specific subjects may be included, such as using a kinesthetic activity to teach revolution and rotation in Earth Science, chemical bonding in Physical Science or Chemistry, or a food web in Life Science or Biology.

Application – In this section, we explain how the technique or activity might be used in one specific class (with the grade or subject identified). The lesson concept is listed and content and language objectives that this technique can help students meet are presented. As needed, key vocabulary terms are identified, particularly when the language objective is related to vocabulary learning. Then, a classroom vignette illustrates how a teacher would use the technique or activity in the lesson.

Differentiation Options – Many of the techniques lend themselves to differentiation. As appropriate, we provide some suggestions for modifications. The adjustments might help teachers apply the technique to students who are at different proficiency levels or grade levels or to students who are underschooled or who have wide gaps in their formal educational backgrounds.

Lesson Preparation

- Alternate Materials
- Enlarged/Adapted Text

The specific techniques described here offer targeting suggestions for preparing and adapting supplementary materials in science lessons. Coupled with strong content and language objectives and meaningful activities, these techniques will help teachers prepare successful SIOP® science lessons.

Alternate Materials (adapted from *99 Ideas*, p. 13)

COMPONENT: Lesson Preparation

Grade Levels: All
Grouping Configurations: Individual, partners, small groups, whole class

Approximate Time Involved: 5–60 minutes
Materials: Real-life items that can help students better understand the content

Description

In all science classrooms, it is necessary to provide students with hands-on activities, images, simulations, videos, and realia. Fortunately, due to the nature of science, these materials are often straightforward and easy to find. Use Figure 2.1 as a guide for how to incorporate these materials into a science classroom.

Grade 7 Physical Science Application

Lesson Concept: States of Matter

Content Objective: Students will be able to match the change in molecular motion when a substance undergoes a phase change to real-life examples.

Language Objective: Students will be able to orally describe the changes in molecular motion when a substance undergoes a phase change.

Key Vocabulary: evaporation, condensation, sublimation, melting, freezing, water vapor, boiling, phase, solid, liquid, gas

In a middle school science classroom, the students are learning about the phases of matter. One of the vocabulary words is *sublimation.* Sublimation occurs when a substance changes from a solid phase to a gas phase. To make this word real for the students, the teacher brings in a sample of dry ice. The students are instantly hooked into the lesson!

The teacher also has other real-life items that undergo phase changes to teach other key terms and concepts, such as ice cream (*melting* and *freezing*), chocolate ice cream topping that becomes frozen when it comes into contact with the ice cream (*freezing*), ice cubes (*melting* and *freezing*), and boiling water (*evaporation, condensation*). These real-life examples help students visualize the different states of matter and make connections with the science vocabulary words.

The teacher then has students watch video clips of substances undergoing phase changes. After each example, she stops the video and has partners discuss what happened and record the type of change in their notebook. After eight clips are shown, she uses the "bump method" to ask the students to describe what occurred using scientific terms. Each student who responds "bumps" the next student metaphorically who will respond (i.e., chooses the next one).

Differentiation

For older or more advanced students, the phase change examples could be set up in hands-on inquiry stations.

For younger students, the phase change examples can be done as a teacher demonstration.

FIGURE 2.1 *Content Ideas for Alternate Materials*

General Area of Science	*Topic*	*Alternate Materials*
Nature of Science	Graphing	Graphs from the Internet, newspaper, or magazines
	Data collection	Images of scientists collecting data from the Internet, newspaper, or magazines
	Science careers	Plan a "Science Career Day" and invite local community scientists and health care professionals to school to do a short presentation about their career
Earth Science	Layers of the Earth	Display a poster of the layers of the Earth on the wall; find Internet images of the layers of the Earth; have students create a model of the layers of the Earth using modeling clay
	Volcanoes	Show photos, images, or videos from the Internet of real volcanoes; have students create a model volcano with labels of volcano anatomy; bring in samples of volcanic rocks or volcanic ash
	Theory of plate tectonics and Pangaea	Show students an Internet simulation of the continents moving apart; show an Internet simulation of the movements at plate boundaries
	Rock cycle	Have samples of many different rocks available for students to observe and classify; have students find a rock from home to bring in to school
Physical Science	States of matter	Post images of the different states of matter on the wall; bring in real examples of matter changing states (dry ice, ice cubes or ice cream melting, water boiling, ice cream topping that turns into a solid when it comes into contact with the ice cream)
	Forces and motion	Show video images of forces and motion—any crash scene in a movie can illustrate some of the concepts; take the students outside to play tug o'war to illustrate the concept of forces and friction
	Electricity	Show images or videos of lightning; use a balloon to illustrate static electricity; allow students to create an electric circuit
Life Science	Human body	Contact a local butcher shop to ask for eyeballs, lungs, and other body parts to show the students; create a life-size poster of a human and label the different parts of the body
	Adaptations	Show images of animals with very clear adaptations—an elephant's trunk, a giraffe's neck, a fish's fins
	Migration	Bring in maps that show the migration patterns of one organism and then have students create their own migration map for another organism; show video clips of animals migrating

Enlarged/Adapted Text

COMPONENT: Lesson Preparation

Grade Levels: 3–12
Grouping Configurations: Individual, partners, small groups, whole class
Approximate Time Involved: Will vary according to length of texts
Materials: Science text that can be enlarged through retyping, overhead projector, or document reader

Description

Science textbooks can often be intimidating and overwhelming for English learners. Adapting the text allows students to focus on the essential information and learner strategies for comprehending the text.

In all classrooms, but particularly heterogeneous classes that include both English learners and native English speakers, it is the teacher's responsibility to provide effective instructional assistance to students who need it. Adapting the text is one way to differentiate the learning process. Once students have an understanding of the information presented in the adapted text, it may be appropriate to extend their learning with the original text.

Steps for adapting the text include:

1. Identify key points of information about living things from a lengthy text that students are expected to read.
2. Rewrite the text, eliminating extraneous information that is not essential to the learning.
3. Retype the text with an enlarged font.
4. Print selected key vocabulary words in bold. Consider bolding fewer words than are presented in the original text.
5. Model for students how to highlight and/or underline important concepts while reading one paragraph at a time.
6. Show younger students and beginning English speakers how to draw pictures on the text's page to reflect vocabulary and conceptual understandings. These can serve as reminders for remembering key information.

Grade 3 Science Application

Lesson Concept: Characteristics of Living Things

Content Objective: Students will be able to identify the six characteristics of living things.

Language Objective: Students will be able to read a passage and determine the main concepts.

Key Vocabulary: cells, energy, grow, reproduce, respond, adapt

(continued)

A third grade teacher prepares a lesson on the six characteristics of living things. She has used the science textbook in previous lessons, but has noticed the text is overwhelming for most of her students and she often ends up reteaching the material. She decides to adapt the text so students can access key information the first time.

After reading the section that focuses on the characteristics of living things, the teacher identifies six key vocabulary words that represent the six characteristics of living things (*cells, energy, grow, reproduce, respond, adapt*). She then goes back through the text and determines which information is essential and which information is extraneous. She retypes the essential information using a larger font and bolding the key vocabulary words. At the bottom of each page, she adds 6 blank lines on which students can note learning right on the page.

When she gives the adapted text to the students, she models how to read, highlight, and note important points. She reads the first paragraph aloud and models her thinking and note-taking. Students read the second paragraph with a partner, highlighting and noting important points. The teacher then randomly calls on students to share their responses. When the teacher sees that students are able to access the information in the adapted text, she has them finish the reading, highlighting, and note-taking with a partner.

To wrap up the activity, the teacher again randomly calls on students to share some of their learning with the whole class. She also collects the notes for review.

Differentiation

For students at newcomer and early beginning proficiency levels, the teacher may need to include many graphics with the adapted text to support the reading.

This activity may also be used in the Strategies component to teach how to determine importance when reading expository text.

Building Background

- Magic Curtain of Science
- Quickwrite
- Oh Yesterday!
- Signal Words
- 4-Corners Vocabulary
- Vocabulary Key Rings

The techniques described next help tap students' background knowledge, build additional knowledge they need for the science work, tie new learning to prior lessons, and develop the scientific vocabulary they need to participate effectively in class.

Magic Curtain of Science

SIOP® **COMPONENT:** Building Background

Grade Levels: All
Grouping Configurations: Whole class

Approximate time involved: 10–15 minutes
Materials: Large piece of fabric and any realia that applies to the topic

Description

The "Magic Curtain of Science" is an activity that allows students to make connections between their lives and science topics that may seem obscure. It is a great activity for an introductory lesson, and students tend to be curious about the Magic Curtain from the moment they walk into the classroom.

Procedures:

1. Place 8–10 items in the front of the classroom on a table underneath a piece of fabric called the "Magic Curtain of Science."
2. Tell the students that you will be lifting the curtain up for just 10 seconds so they can see what is under the curtain.
3. Instruct the students to try to see and remember as many things as they can during the 10 seconds.
4. Lift the "curtain" for 10 seconds and then replace it. After the curtain is replaced, instruct the students to make a list of as many things as they can remember.
5. Then, call on students to share what they saw under the curtain. Make a list on the board of the items the students listed and be sure to add any the students did not notice.
6. Finally, have students predict what they think the lesson might be about.

Other topic ideas for "Magic Curtain of Science" include common items made of chemicals or elements, types of trees (show leaves or seeds), lab equipment, measuring tools, simple machines, and common items made from natural resources.

Earth Science Application

Lesson Concept: Minerals

Content Objective: Students will be able to name 8 items from their daily lives that are made of common minerals.

Language Objective: Students will be able to write a sentence using sentence frames and read it to a partner.

Key Vocabulary: minerals, aluminum, tin, glass, copper, lead, graphite, zinc, iron

In a seventh grade classroom, the students are studying minerals. The teacher arranges the following items under the Magic Curtain of Science: aluminum can, glass beaker, tin can,

(continued)

pencils, silver and/or gold jewelry, copper pennies, light bulb, film, batteries, and a small electronic item.

The teacher conducts the Magic Curtain activity using the steps presented above. After the students' list is recorded on the board, the teacher tells them, "Look! You already know about minerals because you have just listed common items that are made of minerals." The teacher continues the lesson by explaining to the students what type of minerals each item is composed of (e.g., "the soda can is composed of aluminum") and writes the mineral names on the board. She then allows the students to think-pair-share using phrases as sentence frames:

Some examples of real-life items that are made of minerals are ____.

______, _______, and ______ are made of minerals.

Participating in this Magic Curtain of Science activity allows students to connect their new knowledge of minerals with their background information about common items.

Quickwrite (also suitable for Review & Assessment)

COMPONENT: Building Background

Grade Levels: 3–12
Grouping Configurations: Individual
Approximate Time Involved: 5 minutes
Materials: Paper and pencil

Description

A Quickwrite is a simple way to help students remember and connect with prior knowledge. There is no right or wrong answer in a Quickwrite, and it is a safe way for students to share their connections to a certain topic.

Upon direction, students write what they know about a topic for a brief period of time, usually 2–3 minutes. A wide range of topics suit a Quickwrite, including butterflies, volcanoes, earthquakes, stars, natural forces, animals, insects, roller coasters, the solar system, magnetism, electricity, and so forth.

The rules for Quickwrite are

- Spelling and grammar do not matter.
- Students must be writing during the entire time allotted.

Usually after the students have finished writing, they share out, perhaps with a partner first or in a small group. A Quickwrite also gives the teacher insight into students' misconceptions. Uncovering misconceptions is an integral part of science education.

Biology Application

Lesson Concept: Cells

Content Objective: Students will be able to connect what they already know to the lesson topic of cells.

Language Objective: Students will be able to write a sentence using their background knowledge of cells and sentence frames:

One thing I already know about _____ is _______.

"I remember learning ______ in ______."

"_____ reminds me of ______ because ______."

Key Vocabulary: plant cell, animal cell, cell wall, cell membrane, cytoplasm, nucleus

In a high school Biology class, the students are about to begin a unit of study on cells. The teacher gives the students the prompt, "Today during the Quickwrite, you will write everything you know about cells. What are they made of? What do they do? Where are they found?" The teacher makes sure the students understand the procedure's directions and rules. He informs them that they have 3 minutes to complete the Quickwrite in their science notebooks. The students begin writing!

After the Quickwrite, the students pick one of the prompts posted with the language objective to write a complete sentence on their notebook page. They then think-pair-share their connections and background knowledge. In this case, students are able to say that there are different kinds of cells and that they remember looking at cells using a microscope in junior high school.

The teacher uses their background knowledge and connections as a springboard into the unit on cells.

Differentiation

Newcomers and beginners might draw pictures or list words they know that are related to the topic in their notebooks during a Quickwrite. They might also be paired with more advanced students to discuss the topic and then the advanced partner writes.

Oh Yesterday! (suggested by Lindsey Hillyard)

COMPONENT: Building Background

Grade Levels: K–12
Grouping Configurations: Individual
Approximate Time Involved: 5 minutes
Materials: None

Description

Oh Yesterday! is a fun method for students to recall and relate information they learned in the prior lesson. The student selected to begin stands and states something learned or remembered from the lesson the day before, using a theatrical manner, with the sentence starter, "Oh, Yesterday, I learned that . . .". S/he may spread his/her arms wide or take a bow, for example, and speak loudly or in a stage whisper or in a funny voice.

Sentence starters may vary:

Oh Yesterday! I learned that . . .
Oh Yesterday! we studied. . . .
Oh Yesterday! I discovered that . . .
Oh Yesterday! our class . . .

Oh Yesterday! should be done quickly with students speaking one after the other. The teacher may select the speakers or the first student may choose the second, who chooses the third and so on. Usually only 4–6 students do this on any given day.

Because this is an oral activity, students of all ages and language proficiencies can participate.

Grade 6 Science Application

Lesson Concept: States of Matter

Content Objective: Students will be able to review facts about states of matter.

Language Objective: Students will be able to state facts they have learned with inflection and emotion.

Key Vocabulary: liquid, gas, solid, water vapor, ice, melt, freeze, boil, condense

In a sixth grade science class, the students have been studying the three states of matter. The day prior to this lesson, they conducted some experiments with water: melting ice, freezing water, and boiling water with and without a lid on the pot.

At the start of this lesson, the teacher used the name sticks to select the first student to perform Oh Yesterday! (A name stick is a popsicle-like stick with a student's name written on the side. A class set may be kept in a cup and used to call a student at random, by pulling out a stick.)

The student stood and shouted "Oh Yesterday! we changed water into ice." He pulled a name stick to choose the next speaker. She stood and bowed while saying softly, "Oh Yesterday! we discovered water vapor could become water again when it cools." She then pulled another name stick and this process continued for 2 more students.

Differentiation

The teacher might provide rehearsal time before having students perform. Newcomers and beginners might be paired up with other students to generate Oh Yesterday! sentences and practice before being called on.

More advanced students might be challenged to make an analytical statement or draw a conclusion about the prior lesson.

Signal Words (Adapted from Sarah Russell's contribution in *99 Ideas*, p. 36)

SIOP® COMPONENT: Building Background

Grade Levels: 3–12
Grouping Configurations: Independent writing and reading
Approximate Time Involved: 10 minutes
Materials: Chart paper and markers for Signal Words posters

Description

Various types of text structures are found in science textbooks and required for writing assignments. For example, students may be asked to compare/contrast, sequence, describe, or explain cause and effect. Teachers can prepare a "Signal Words" poster to help students recognize and understand these structures. The posters would list words and phrases to signal the types of text structure that students may be working with. Posters can be displayed around the room for students to refer to as they read and write.

These posters are useful for building background schema. When students encounter these phrases, for instance, they have an initial idea of what type of text they will be reading, so they will have a jump start on comprehension. The following is a sample poster.

Signal Words

If you are asked to write a

DESCRIPTION OR LIST

use these words:

To illustrate . . .
For instance . . .
In addition . . .
And . . .
Again . . .
Moreover . . .
Also . . .
Too . . .
Furthermore . . .
Another . . .
First of all . . .

Grade 5 Science Application

Lesson Concept: Transpiration

Content Objective: Students will be able to explain the steps of transpiration.

Language Objective: Students will be able to use key vocabulary to write a paragraph describing transpiration.

(continued)

Key Vocabulary: evaporation, soil, water, nutrients, stomatas

A class is learning about the steps of transpiration. The teacher is also teaching students to write about expository text and is focusing on how to write the steps in a sequence. For this particular lesson, students are asked to read from the science text and use small self-stick notes (one self-stick note can be cut into smaller pieces) to identify the steps in transpiration.

The teacher begins by teaching students the key vocabulary and then introducing students to a new "Signal Words" poster that is in the room. (See poster below.) Students then read the text, using the self-stick notes. They are reminded to look for key vocabulary and words from the "Signal Words" poster.

Signal Words

If you are asked to describe the

Sequence or Order

use these words:

first
second
third
then
next
after that
finally
first of all
before
meanwhile
last
now
later on

The teacher then models how to use the key vocabulary words and words from the "Signal Words" poster to write a summary paragraph. The teacher models the first sentence. For the second sentence, the teacher asks students to come up with a sentence with their partner and then share out. When the teacher sees that students are using both key vocabulary and words from the "Signal Words" poster, students work with a partner to finish their paragraphs.

Differentiation

It is useful to introduce one poster, one text structure, and one set of signal words at a time.

For students with lower levels of English proficiency, include only 2–3 signal words at first. Give the students time to learn and apply them before adding new words.

For students at advanced proficiency levels, also include only 2–3 words at first, but encourage the students to add new words as they encounter them.

4-Corners Vocabulary (Adapted from *99 Ideas*, p. 40)

SIOP® **COMPONENT:** Building Background

Grade Levels: All
Grouping Configurations: Individual, partners, small groups, whole class
Approximate Time Involved: 15–30 minutes
Materials: Chart paper and markers, or paper and pencil for upper grades

Description

The purpose of 4-Corners Vocabulary is to allow students to learn a vocabulary word from different perspectives. Using a piece of paper or chart paper folded length-wise and width-wise to make four boxes, students create a 4-Corners Vocabulary chart. In each respective box they draw an illustration of the word, write a definition of the word, write a sentence using the word in context, and list the actual vocabulary word.

FIGURE 2.2 *Sample Vocabulary Topics for 4-Corners Vocabulary*

General Science	*Earth Science*	*Physical Science*	*Life Science*
Types of graphs and data tables	Landforms	Phases of matter	Types of tissue
Lab equipment	Types of volcanoes	Simple machines	Plant anatomy
	Components of the water cycle	Chemical bonds	Flower anatomy
	Phases of the Moon	Molecules	Stages of a life cycle
	Celestial bodies	Atoms	Cell organelles
	Types of clouds	Atomic particles	Types of cells
	Fault lines		Bones of the body
			Body parts
			Organs
			Types of animals
			Phases of mitosis
			Types of seeds

Life Science Application

Lesson Concept: Population and Ecosystems

Content Objective: Students will be able to draw an illustration to show their understanding of a scientific term.

Language Objective: Students will be able to write a definition and contextualized sentence to show understanding of a vocabulary word.

(continued)

Key Vocabulary: population, emigration, immigration, exponential growth, abiotic, biotic, predator, prey, ecosystem, carrying capacity, limiting factor

The teacher does an example with the class using the word *population.* As she creates the example, she explains the steps. The final 4-Corners Vocabulary Chart looks like Figure 2.3.

FIGURE 2.3 *4-Corners Vocabulary Chart for "Population"*

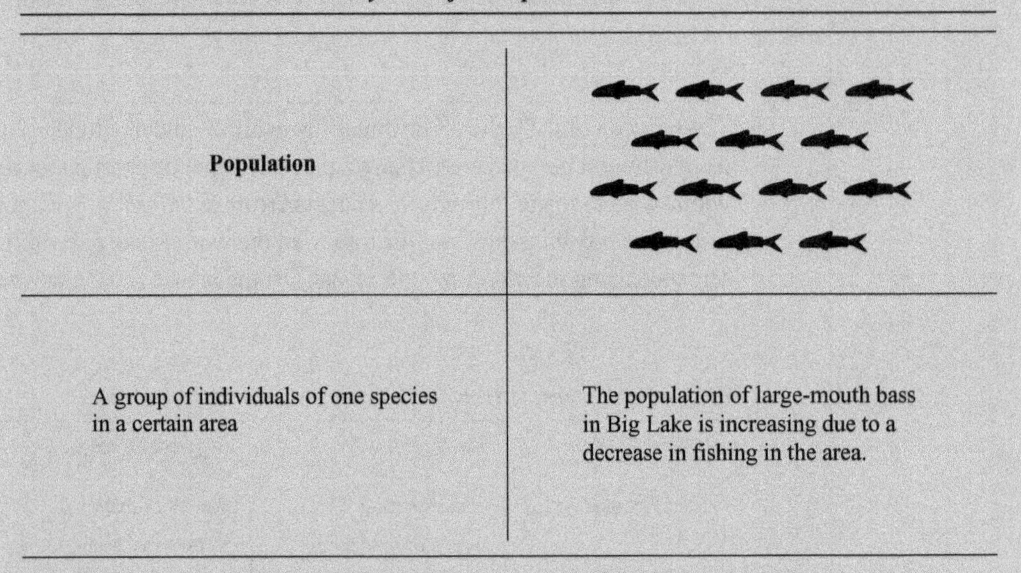

Next, the teacher gives other vocabulary words for the unit such as *emigration*, *immigration*, *predator*, and *prey* and asks the students to create 4-Corners Vocabulary Charts for them. Students may work individually, with partners, or in small groups. They draw an image that helps them connect with the word, write a sentence using the word, and write a definition in their 4-Corners Vocabulary Charts.

Differentiation

Teachers may assign words and groupings according to student proficiency levels. Some students may work independently on 3 words apiece, some may work with partners. Lower level students may work in a group with more advanced students.

If desired, these charts can be shared in a game. Students fold their papers with only the illustration showing. They form two teams. Each team takes turns showing the illustration to the other team. If that team guesses the word, it gets a point. After a round with illustrations, the teams refold any charts with words that haven't been guessed, so the sentence is showing. The game continues. If some words are still not guessed at the end of this second round, the remaining charts can be refolded once more with the definition showing and the game continues. After this round, the team with more points wins.

These charts are excellent resources for word walls.

Vocabulary Key Rings

COMPONENT: Building Background

Grade Levels: K–6
Grouping Configurations: Individual
Approximate Time Involved: 10 minutes
Materials: Key rings, index cards, hole punch

Description

Students are introduced to key vocabulary words for the unit. The introduction will include a student-friendly definition, graphics, realia, and other input to make the terms comprehensible. Students are then given note-cards to represent each word (one card per vocabulary word). On the front of the card, students write the word. On the back, students draw a picture of the word. (For older students, the back of the card may also include the definition.) When the teacher has reviewed the cards with the students and clarified any misconceptions, cards are then hole punched in the upper-left-hand corner and added to that student's "Vocabulary Key Ring." Students can have their "Vocabulary Key Rings" with them when they are discussing, reading about, or writing about a topic.

Kindergarten Science Application

Lesson Concept: The Life Cycle of the Butterfly

Content Objective: Students will be able to distinguish among the stages in the life cycle of the butterfly.

Language Objective: Students will be able to use key vocabulary and sentence starters to describe the life cycle of the butterfly to a partner.

First there is a ___________.
The ________ turns into a ________.

Key Vocabulary: cocoon, chrysalis, butterfly, caterpillar, larva, life cycle

Using teacher-selected materials, the teacher introduces the stages in the life cycle of the butterfly to the whole class. (Materials could include posters, charts, literature books, etc.) The teacher asks the students what they already know about butterflies. (Have you ever seen a butterfly? Have you seen a caterpillar? What do you know about butterflies? etc.) As the teacher introduces each stage, he/she models drawing a picture to represent the stage and writing the name of the stage on an index card. The teacher also models and allows students to create cards for any related vocabulary words (e.g., life cycle). Students are then given index cards and will create their own drawing and write the name of

(continued)

the stage on their own index card. (The teacher may want to hole punch the cards before handing them out to students.) As students create their cards, they are distinguishing among the different stages.

The teacher calls a student to be his/her partner and models how to use the sentence starters to discuss the learning. Students then are assigned a partner and use the provided sentence starters to discuss what they learned. As students talk with their partners, the teacher circulates the classroom to check for understanding and monitor language use.

The teacher then calls on five students to each represent one stage. Out of order, she asks them to pantomime the stage. Classmates use sentence starters to explain which stage they are seeing demonstrated.

When students are finished, they add their index cards to a key vocabulary ring. As vocabulary words are added, the students can continue to refer to them as they proceed through content.

Differentiation

Advanced/Transitional: For students with strong academic English proficiency, the teacher may decide on a wider variety of sentence starters. For example:

The cycle begins with a ___________.
The second stage is __________.
Finally ________ will become_________.

Comprehensible Input

- Procedures with Graphics
- Framed Outlines
- *Move It!*

The techniques described here give teachers ideas for helping students access the scientific information needed in their lessons. In many cases, students benefit when the information is presented in smaller chunks or steps and then pulled together. Familiar ESL techniques such as using visuals, demonstrations, gestures, movement, role plays, and simulations are very applicable to the science classroom.

Procedures with Graphics

COMPONENT: Comprehensible Input

Grade Levels: All
Grouping Configurations: Individual, partners, small group, whole class
Approximate Time Involved: 5 minutes
Materials: Markers and chart paper, dry erase board, or overhead projector

Description

Written procedures with graphics help make content more comprehensible to students. In science, it is especially important for students to be able to follow procedures in experiments.

When explaining an experiment or other activity, the teacher states the directions, shows the students directions that are displayed on the board, shows students the images that align with the directions, and, finally, models the directions. By including each of these aspects to communicate the directions, the teacher gives students many opportunities to comprehend the directions before they carry out the activity. After delivering the directions, the teacher can instruct the students to think-pair-share the directions and then call on a few students to repeat or paraphrase for the class.

Physical Science Application

Lesson Concept: Chemical Reactions

Content Objective: Students will be able to follow written procedures for an experiment.

Language Objective: Students will be able to use sequence words to rewrite the steps in the experiment.

Key Vocabulary: scoop, swirl, first, second, third, next, then, after that, finally

In a middle school Physical Science classroom, the students are learning about chemical reactions. The students are going to participate in a simple activity that lets them observe a chemical reaction first-hand. Although the procedure seems simple, it is important for the teacher to have very clear directions and expectations for the students. This can be accomplished by using procedures with graphics. First, the teacher asks the students to read the procedures silently as she reads them aloud to the class. Next, she demonstrates the procedures as she reads them aloud again. At this point, she checks on student knowledge of sequence words, such as *first*, *next*, *after that*, and calls their attention to a Signal Words poster containing sequence words. She then has the students write the directions on their paper using the sequence words. Finally, the students tell a partner the steps to the directions using sequence words.

After the teacher calls on a few students to share their sequenced steps with the class, the teacher will be able to assess whether the students understand the procedures. If the students do understand, they will proceed with the experiment. If the students do not seem to understand, the teacher might decide to do the experiment as a class demonstration, one step at a time.

(continued)

See example procedures with graphics below:

FIGURE 2.4 *Sample Directions for Procedures with Graphics*

Directions for Penny Experiment

1. Get supplies. salt penny small beaker vinegar spoon
2. Put one scoop of salt into the beaker.
3. Put the penny in the beaker.
4. Add about 20 ml of vinegar to the beaker.
5. Swirl around for two minutes.
6. Remove the penny and record your observations.
7. Clean up your area and be prepared to discuss your observations. what 1¢ happened

Differentiation

Depending on the proficiency level of the students and their prior experience doing science experiments, the teacher might deliver instructions step-by-step and wait for the students to do each step before moving on to the next one. For example, for the penny experiment described here, the teacher would first have students get their supplies. When all of the students have their supplies, the teacher will read the second step. Then, as a class, all of the students will add the scoop of salt to the beaker. When all students are done with this step, the teacher will read the directions to the next step. The teacher will continue reading and demonstrating each step until the students have completed the experiment.

Framed Outlines

COMPONENT: Comprehensible Input

Grade Levels: All
Grouping Configurations: Individual
Approximate Time Involved: 5–20 minutes, depending on length of clip, passage, or lecture
Materials: Outline of lesson content with key information excluded

Description

Framed outlines are designed to scaffold student comprehension of a piece of text, a lecture, or a lesson. The teacher creates an outline with some key information intentionally left out.

Students can then fill in the outline as they participate in a lesson, listen to a mini-lecture, read a passage, or watch a video clip. Students can refer back to the outlines in future lessons. They are particularly useful for English learners because they provide a focus and a structure for students to organize their thoughts while supporting them with the academic language load.

Grade 1 Science Application

Lesson Concept: Sonoran Desert Habitat

Content Objective: Students will be able to identify plants and animals found in the Sonoran Desert.

Language Objective: Students will be able to view and listen to a video clip in order to record key information about a habitat.

Key Vocabulary: coyote, jack rabbits, King snake, cactus, cactus wren, succulents

Students study the Sonoran Desert and the animals that live there. Students view a video clip about plant and animal life in the Sonoran Desert. From time to time as the video plays, the teacher pauses it and allows students to discuss key points. Students add information to their Framed Outlines (see below). When the video clip is finished, students review their Framed Outlines by reading through their outline with a partner and comparing answers. If two partners have different answers, they may consult with another student or the teacher.

Framed Outline:

The Sonoran Desert is filled with many plants and animals. One animal that lives in the Sonoran Desert is a ________________. It lives in a ________________ and it eats ________________. Another animal found in the Sonoran Desert is a ________________. It is ________________ and ________________. Some of the plants you would find in the Sonoran Desert are ____________________ and ____________________.

Differentiation

Beginning/Early Intermediate: For students at the beginning/early intermediate levels, the teacher may need to support learning from the video clip with additional visuals or explanations. For example, rather than simply pausing the video tape and asking students to fill in their outlines, the teacher might pause the video, offer a simple summary, and model filling in the Framed Outline.

Move It! (from 99 Ideas, p. 53)

COMPONENT: Comprehensible Input

Grade Levels: All
Grouping Configurations: Individual, partners, small group, whole class
Approximate Time Involved: 5–30 minutes
Materials: None

Description

In *Move It!* students use facial expressions, hand gestures, or body movements to learn academic content. The physical movements afford students an opportunity to discover vocabulary and science processes using kinesthetic learning.

The following chart offers *Move It!* ideas for a variety of scientific topics.

FIGURE 2.5 *Science Topic Ideas for* Move It!

General Science	*Earth Science*	*Physical Science*	*Life Science*
Graphing (students create a human bar graph, a line graph, or a scatter plot)	Plate boundaries (students use hand motions to show plate movement and the resulting landforms) Revolution and rotation (students act out the different movements in small groups)	States of matter (students act out the movements of molecules in each state) Chemical bonding (students create life size molecules by joining with other student "atoms") Inertia (students experience inertia firsthand by running at full speed and trying to stop when the whistle blows)	Types of animals (students use arms to act out elephant trunk or alligator jaws, and so on) Life cycles (students use hand gestures for caterpillar, cocoon, butterfly) Photosynthesis (students act out the process of photosynthesis)

Earth Science Application

Lesson Concept: The Theory of Plate Tectonics

Content Objective: Students will be able to identify four types of plate boundaries.

Language Objective: Students will be able to express if–then statements to a partner about the different plate boundaries.

(continued)

Key Vocabulary: plate tectonics, boundary, transform, divergent, collision, convergent, if, then

While studying the Theory of Plate Tectonics, students learn about the different plate boundaries, such as transform, divergent, collision, and convergent boundaries. For language learners, these concepts are difficult and hard to remember. In order to help students learn and retain the concepts, the teacher demonstrates hand motions and then has the students practice the hand motions while saying the words.

For example, to demonstrate a collision boundary, students hold arms out in front, bent at the elbows, about shoulder high with palms facing down and fingers on both hands pointing toward the middle, but apart. The students then bring their hands together until their fingers are touching and push them up into the air to show that the plates collide and create mountain ranges.

After the students have mastered the motions and key terms, the teacher introduces if–then statements, such as "If the plates collide, they create mountain ranges." She has them use the hand motions and state the if–then statements as she teaches them. She then asks students to volunteer to demonstrate for their classmates. She calls out a type of boundary, such as divergent boundary, and that volunteer demonstrates the motion and expresses the if–then statement.

Differentiation

For students with lower language abilities, the teacher will create and model the *Move It!* techniques. For students with higher language abilities, the students can create their own motions for the topic.

Strategies

- SQP2RS (Squeepers)
- The Insert Method
- Stop and Think
- T-Chart

By using the techniques described here, teachers can help students develop facility with metacognitive and cognitive strategies. They can be applied to tasks requiring listening and reading comprehension.

SQP2RS (Squeepers) (Adapted from *99 Ideas*, p. 71)

COMPONENT: Strategies

Grade Levels: All
Grouping Configurations: Whole class, small groups, partners
Approximate Time Involved: Varies according to length of text and age and proficiency levels of students
Materials: Informational or expository text, overhead projector or document reader

Description

SQP2RS ("Squeepers") is an instructional framework designed to support students as they read expository or informational text. (SQP2RS is not designed to be used with narrative text, poetry, or stories.) The framework scaffolds the metacognitive strategies that proficient readers use (e.g., predicting, self-questioning, monitoring/clarifying, evaluating, summarizing/synthesizing). The practice of interacting with text in this way is very valuable. Students can learn what to do when they are faced with challenging texts if strategic thinking and reading are taught and reinforced. When teaching SQP2RS it is important to model each step and scaffold the process so students advance toward independent use of the technique. To model the activity, an overhead or document reader could be used. The six steps to the SQP2RS instructional framework are:

Survey: Students are given 1–2 minutes to skim the assigned text or pages of a chapter. Students may be instructed to look for words in bold print, italics, graphs, pictures, and so on. Model your own thinking aloud first to teach students to survey the assigned section of text.

Question: During this step, student groups formulate questions about the text that they have just surveyed. Limit the number of questions to 3 or 4. Instruct students to create questions they believe they will find the answers to as they read the text. Students may then share their questions with the whole group. As the students pose the questions, the teacher records them. If a question is repeated by another group, the teacher puts a "star" next to it, indicating that it may be an "important" question because more than one group thought of it.

Predict: Next, the groups generate predictions of what they think the text will address. By doing this, students determine what they think the important concepts in the text will be. Determining the major concepts and ideas can be the most challenging part of reading expository text. The teacher's responsibility is to help students narrow their focus, using the questions to predict four or five concepts that they think the text will focus on. The predictions may even restate some of the questions. Predicting and questioning are integrated thinking processes.

(Once students have learned and practiced the process, the first three steps of Squeepers (Survey, Question, Predict) shouldn't take more than 5–7 minutes.)

Read: Students read from the expository/informational text in pairs or small groups. (While reading, students note answers to the questions they have formulated about the topic or places where their predictions have been confirmed. Writing answers on self-stick notes and marking the text where the answer is found is a good technique for getting students to return to the text and find evidence for their responses.)

Respond: Students discuss the information they have gathered in the first four steps of Squeepers. They also discuss any questions that they did not find answers to in the text. The response time is designed to provide students with opportunities to interact with one another and discuss their learning. This opportunity acts as a scaffold for them to complete a more structured summary.

Summarize: Students summarize key concepts learned from the text. They may summarize orally or in writing and they can work individually, with a partner, or in a group. Students should use key vocabulary in their summaries when appropriate.

Grade 4 Science Application

Lesson Concept: Erosion, Deposition, and Weathering

Content Objective: Students will be able to explain the role that erosion, deposition, and weathering plays in altering the Earth's surface features.

Language Objective: Students will be able to monitor their reading comprehension by formulating questions and predictions, responding, and confirming predictions.

Key Vocabulary: erosion, deposition, weathering, surface, alter

To help students activate and use their background knowledge and experience related to the Earth's surface features, the teacher begins the SQP2RS activity by modeling her thinking aloud during the survey step. She distributes copies of five pages from the chapter in the school science textbook and also displays the text on an overhead transparency. As she teaches the class to survey, she says, "I see that the word *erosion* is next to a picture of a river bed. The word *deposition* is next to a picture of the beach. I wonder what these mean? I also see that the words *erosion*, *deposition*, and *weathering* are in bold." Students then spend 2 minutes surveying the text with a partner.

Next, the teacher tells students they will work with their partners to create 3–4 questions they believe they will find the answers to in the text. The teacher models writing a question and offers sentence frames for students to use (e.g., How does _________ affect the Earth? What does _________ cause?). Students then share their questions and the teacher records the questions on the transparency.

The teacher then reminds students that the content objective is to explain the role that erosion, deposition, and weathering play in altering the Earth's surface features and their predictions should focus on that. Students work with their partners to make predictions about what they will learn. The students write their predictions in their science notebooks. The teacher spends 1–2 minutes asking students to share one of their predictions with the whole class.

When the class has shared some predictions, the teacher models reading the section of the text. (The section should be short, but ideally include an answer to a question that was asked.) The teacher models how to record a response on a self-stick note, then instructs students to read the text with their partner. As students read, they note any answers they find on self-stick notes.

The teacher brings the class back together as a whole group. She asks students to use sentence starters to discuss the answers they found. The teacher records the responses next to the questions on the transparency as they answer. Students write the answers in their science journals. For some unanswered questions, the teacher points out, "We might find answers to some of the questions in the next section or another chapter." She adds, "Did you think of new questions while reading? Are there questions we should add or eliminate before we continue reading?"

After using the sentence starters to respond to the reading, students are prepared to summarize the key concepts they learned about the role that water plays in processes that alter the Earth's surface features. The teacher reminds students that their summaries

(continued)

Grade 4 Science Application *(continued)*

should include the terms *erosion, deposition*, and *weathering*. As needed, students can use the sentence starters to write their summaries. The teacher can determine whether the summaries should be oral or written, and if they are written, what the format should be.

Differentiation

Kindergarten and Grade 1 teachers use informational big books; Grades 3–12 teachers use informational or expository texts.

Once students are familiar with Squeepers, the teacher may decide to have them read with partners or independently.

Graphic organizers may be provided for writing questions, answers, and predictions.

Students might also choose to highlight or write their notes directly on the text, if there are individual copies available.

Students can use sentence starters to summarize their learning.

I learned . . .
I noticed . . .
I was surprised that . . .
________ causes ______
I did not find out if . . .
I still need the answer about . . .
If _____ happens, then _______

(Sentence starters can be posted for students, or they can be written on note cards and placed on key rings for students to use individually.)

The Insert Method (Adapted from *99 Ideas*, p. 33)

COMPONENT: Strategies (also suitable for Building Background)

Grade Levels: 3–12
Grouping Configurations: Partners, small groups, whole class, individual
Approximate Time Involved: 15–30 minutes, depending on the length of the text
Materials: Informational or expository text, self-stick notes (optional)

Description

Students work with partners to read a nonfiction article or a science textbook using the following coding system. Students can insert the codes directly onto the page, or they can use self-stick notes either with the codes written on them or by writing the codes on them.

When students have finished marking the text, they can move from partners to groups of four to share their questions and new learning with other students. As questions are answered or misconceptions are cleared up, the question marks can be replaced with an asterisk (*). Following the small group work, the text can be discussed with the teacher and/or the whole class.

✓ A check mark indicates a concept or fact that is already known by the students.
? A question mark indicates a concept or a fact that is confusing or not understood.
! An exclamation mark indicates something that is new, unusual, or surprising.
+ A plus sign indicates an idea or concept that is new to the reader.
* An asterisk indicates a question has been answered.

Grade 4 Science Application

Lesson Concept: Biomes of the World (In this particular lesson, students are focusing on shorelines.)

Content Objective: Students will be able to read and identify facts in the text about shorelines.

Language Objective: Students will be able to use the following sentence starters to discuss shorelines and ask questions about shorelines.

I wonder if . . .

What did you understand about . . . ?

I learned . . .

I discovered . . .

Key Vocabulary: evaporation, soil, water, nutrients, dune, shoreline

In this lesson, students work with a partner to read a nonfiction text about shorelines. As they read, they use the symbols written on self-stick notes to identify facts. (The facts may be already known, confusing, new, unusual, or surprising.)

Pairs then team up to create a group of four. In groups of four, students use their self-stick notes and the sentence starters to discuss their questions and their comprehension. At this point, students may replace their question marks with an asterisk if a question is answered or a confusing point is clarified. Next, groups share one or two facts about shorelines with the class. If unanswered questions remain, students pose those questions to the class. Depending on the level of comprehension and the number of questions shared, the teacher may decide to address the questions at the time they are being asked or he/she may record them to address in a subsequent lesson.

(continued)

Grade 4 Science Application *(continued)*

Differentiation

Beginning/Early Intermediate: In a heterogeneous classroom, the teacher may form a small group with beginning/early intermediate level students and read with them rather than asking them to read with partners. If a teacher has a class of students primarily at the beginning/early intermediate levels, he or she may decide to read the text aloud to the whole group, pausing and allowing students time to note and discuss their learning. If some students share the same primary language, the teacher can allow time for clarification in L1.

Advanced/Transitional: For students at these levels, the sentence starters may be unnecessary. The sentence starters may be provided, but students could choose to use other academic language expressions to share their ideas. The teacher may encourage them to combine ideas into compound sentences such as "Many wading birds live in salt marshes while flying birds live on barrier islands."

Stop and Think (Adapted from Shelley Frei's contribution in *99 Ideas*, p. 90)

COMPONENT: Strategies

Grade Levels: All
Grouping Configurations: Individual, partners, small groups, whole class
Approximate Time Involved: 5 minutes
Materials: Informational or expository text, self-stick notes (optional)

Description

As proficient readers work through expository text, they employ many different reading strategies. These strategies may include making predictions, making inferences, visualizing, and creating questions. Students who are working toward becoming proficient readers will need explicit modeling and guided practice of these strategies before they begin to employ them independently. One scaffold teachers can employ to enable students to use these strategies independently is the "Stop and Think" technique.

The teacher demonstrates the technique by reading aloud a few paragraphs at a time, stopping and making connections for the students. The teacher may use sentence starters such as the following to teach students how to articulate their thoughts while using the strategy of making connections:

The ________ reminds me of_____________ because . . .

I have seen a _______. That is similar to. . . .

When I read__________ it reminded me of . . .

Grade 1 Science Application

Lesson Concept: Prairie Habitat

In a first grade class, students are studying different habitats. This particular lesson focuses on the prairie. Students are also practicing the reading strategy of making self-to-text connections.

Content Objective: Students will be able to identify facts about the prairie as they read expository text.

Language Objective: Students will be able to make connections between what they have seen or experienced and the prairie habitat as they read expository text.

Key Vocabulary: grasslands, sediment, glacier, topsoil, prairie, prairie dog, ground hog

The teacher and students each have a photocopied piece of text. (Depending on the materials available, this may be a science textbook, an article, or other adapted text. The teacher may also consider adapting this text.) The teacher reads the first two paragraphs of the science text aloud. As the teacher reads, he stops to model his thinking when making a self-to-text connection. As the teacher models, he may say things like "I have seen a groundhog. It is similar to a prairie dog." or "When I read that there is a lot of topsoil, it reminded me of my garden at home." Each time the teacher stops for a "think-aloud," he can place a finger at the side of his head to demonstrate that he is taking time to "Stop and Think."

Next, the teacher reads the third and fourth paragraphs of the science text aloud, pausing at predetermined places in the text. When he pauses, the students "Stop and Think" about their own self-to-text connections. Students place their fingers next to the sides of their heads and share their connections with a partner. (Remind students to use the key vocabulary and the sentence starters as they share.)

The teacher next calls on students to share their connections with the whole group before continuing to read the text.

Depending on the level of difficulty of the text, the teacher may then decide to have students finish reading by themselves and practice using the "Stop and Think" strategy with a partner.

Differentiation

Newcomer students might benefit from the "Stop and Think" technique to introduce a picture walk or a chapter walk. The teacher would model how to interpret photos, diagrams, charts, headings, and so forth.

Advanced students might apply multiple strategies as they read, but for less proficient students and underschooled students, we need to teach these strategies one at a time and give them plenty of practice.

T-Chart (Adapted from *99 Ideas*, p. 86)

COMPONENT: Strategies (also suitable for Practice & Application, and Review & Assessment)

Grade Levels: All
Grouping Configurations: Individual, partners, small groups, whole-class brainstorming
Approximate Time Involved: 10–25 minutes
Materials: T-Chart to display for whole class, T-Chart graphic organizer for students

Description

A T-Chart is used to help students classify information into two categories. It can be used when students are taking notes about two topics, comparing some concepts, or brainstorming ideas (e.g., pros and cons of a science controversy). A T-Chart can be expanded to include three columns; it is then called an M-Chart.

Teachers can introduce students to a T-chart using a familiar topic, such as comparing rap music to rock-n-roll or listing the pros and cons of studying for a test. Using a large chart, the teacher can ask students to brainstorm ideas for each column and record them for the class to see. After several ideas are listed, the students can work with a partner to finish the chart.

Applications of the T-chart technique can be found in Figure 2.6.

FIGURE 2.6 *Sample Science Topics for T-Charts*

General Science	*Earth Science*	*Physical Science, Chemistry, Physics*	*Life Science, Biology*
Metric units, English units	Sedimentary, igneous, and metamorphic rocks	Kinetic energy, potential energy	Plant cells, animal cells
Safe lab techniques, unsafe lab techniques	Comets, asteroids, meteoroids	Pros and cons of nuclear energy	Plants, animals
Observations, inferences	Inner planets, outer planets	Elements, compounds, mixtures	Dominant traits, recessive traits
	Polar climates, temperate climates, tropical climates	Organic compounds, inorganic compounds	Tissues, organs, systems
		Positively charged ions, negatively charged ions	Vertebrates, invertebrates
		Acids, bases	Mitosis, meiosis

Earth Science Application

Lesson Concept: Nuclear Energy

Content Objective: Students will be able to explain three pros and three cons of nuclear energy.

Language Objective: Students will be able to write an opinion about nuclear energy.

Key Vocabulary: nuclear, energy, radiation, resources, renewable, nonrenewable

In a high school Earth Science class, the students are learning about renewable and nonrenewable energy sources. In this lesson, the students are going to learn about nuclear energy. The students will take notes on a nuclear energy reading. To help them organize their thoughts, the teacher provides them with a T-Chart. The teacher asks the students to read the passage independently and record pros and cons in their charts. They may pair up with a partner to share ideas after they have listed some themselves.

After writing the pros and cons of the issue on their T-Charts, the teacher asks students to think about their opinion on the use of nuclear energy. She has them complete this sentence starter:

I support/do not support the use of nuclear energy because _______.

Differentiation

The teacher may form a small group with younger or less proficient students. The teacher may read the text aloud or support shared reading by the students. As they discuss the pros and cons, the teacher may write on the T-Chart.

Another option is for students with lower language levels to draw images on the T-Chart instead of writing notes.

The opinion sentence starter can be used by students of varying proficiency levels. Advanced students might write a paragraph in response, whereas beginner and intermediate students might write one sentence.

Concluding Thoughts

This chapter is the first of two that offer a selection of activities and techniques that can help teachers meet the goals of SIOP® science instruction. We encourage you to try these activities with your students. Although we have situated them in a particular grade with a particular lesson topic, you will find that these techniques can be applied in multiple contexts and can be differentiated for the range of learners in your classroom. These techniques should not be treated as a grab bag of activities, however. Rather, select from them carefully in service of your lesson goals, your language and content objectives, and the needs and interests of your students.

chapter 3

Activities and Techniques for SIOP® Science Lessons: Interaction, Practice & Application, Lesson Delivery, Review & Assessment

By Hope Austin-Phillips, Amy Ditton, and Deborah Short

Introduction

This chapter presents additional techniques and activities for science instruction, organized by the final four SIOP® components: Interaction, Practice & Application, Lesson Delivery, Review & Assessment. As we did in Chapter 2, we describe the steps for each

technique, contextualize it in a particular grade and science lesson topic, and offer suggestions for differentiation. (If you haven't read Chapter 2 yet, look over the beginning of the section of Science Techniques and Activities, pages 23-51, to see the explanations and purposes of each section of a technique's description.) The techniques are generally useful across the range of K–12 grade levels, although some are more advantageous in a narrower range. You will notice again that some of the techniques are applicable to more than one SIOP® component. Remember, the activities alone do not constitute a lesson. Choose among these proven techniques so students will explore, practice, and meet your content and language objectives.

Interaction

- You Are There
- Milling to Music

The techniques described here offer multiple opportunities for students to interact with each other. By providing sentence starters and academic frames, the teacher can scaffold the appropriate language for the students to use. Students also interact with a variety of texts.

You Are There (Adapted from *99 Ideas*, p. 119)

COMPONENT: Interaction

Grade Levels: 4–12
Grouping Configurations: Small groups, whole class
Approximate Time Involved: 30–45 minutes
Materials: Resources for research (e.g., books, articles, websites)

Description

The teacher divides students into small groups of three or four. Each group is assigned an event to research (historical or current). Students are given time to research their assigned topic.

When the research is completed, students prepare to present the information through an interview activity. They designate roles within their groups. One student plays the role of the interviewer and the others take on the roles of characters, people, or animals that were present when the event took place. For example, one student may portray a cat that was under the porch when a tornado hit, while another student portrays the homeowner who couldn't find his/her cat. The "interviewer" will prepare questions to ask the "interviewees," and the "interviewees" will prepare their responses. The teacher should remind students to use the key vocabulary as well as the provided sentence starters when preparing their presentations.

Next, small groups will take turns acting out their "interviews" for the rest of the class. As the students watch the interviews, they can use prepared outlines to take notes on the information presented. The prepared outlines may also include the sentence starters.

Grade 6 Science Application

Lesson Concept: Natural Hazards

Content Objective: Students will be able to conduct research on natural hazards.

Language Objective: Students will be able to write interview questions and response sentences in order to participate in interviews about natural hazards.

Key Vocabulary: impact, interactions, populations, hazards, environment, sandstorms, hurricanes, tornadoes, ultraviolet light, lightning-caused fires

The class has been studying the effects of the following natural hazards: sandstorms, hurricanes, tornadoes, ultraviolet light, and lightning-caused fires. The teacher divides students into small groups of three or four. Each group is assigned to research a hazard and its effects. The teacher provides printed articles and a list of websites for students to use when conducting their research. Students are given time to research their assigned topic. (For this lesson, students conducted research on day one and prepared their interview questions and responses on day two.)

For the next step, student groups prepare their interviews. The teacher models the question–answer format and reminds students to use the key vocabulary as well. The groups record their interview questions and answers on their outlines (see Figure 3.1).

Next, small groups take turns acting out their "interviews" for the rest of the class. As the students watch the interviews, they use the prepared outlines to help them take notes on the information presented.

Differentiation

The interview could be presented as a newscast, with a reporter interviewing witnesses at a scene, or it could be presented as a talk show with experts joining the host.

Planning time for rehearsal is beneficial for students at lower levels.

FIGURE 3.1 *Natural Hazards Interviews*

Sandstorms

Questions:	*Responses:*

Hurricanes

Questions:	*Responses:*

(continued)

Tornadoes

Questions:	*Responses:*

Ultraviolet Light

Questions:	*Responses:*

Lightning-Caused Fires

Questions:	*Responses:*

Milling to Music

COMPONENT: Interaction (also suitable for Building Background, Practice & Application, and Review & Assessment)

Grade Levels: All
Grouping Configurations: Partners
Approximate Time Involved: 5–10 minutes
Materials: Music and speakers loud enough for the class to hear

Description

Milling to Music is an interaction technique that allows students to have access to and work with multiple partners in a short amount of time. Milling to Music also greatly appeals to musical and kinesthetic learners and gets all students up and moving around the classroom.

This activity can be used to review a homework assignment, class notes, or vocabulary words. Milling to Music gives students access to the ideas of many of their classmates. In this activity, students who are unsure of the response to a prompt (e.g., the definition for a word) are able to exchange ideas with other students in the classroom and improve their own responses.

Procedures:

1. All students must have their assignment, notes, or other material to be discussed in their hands.
2. When the music starts, all students walk around the classroom.
3. When the music stops, each student selects a partner nearby and waits quietly for the prompt from the teacher.

4. After hearing the prompt, the students take turns sharing their responses with their partner.
5. When the music starts again, all students walk around the classroom.
6. When the music stops, each student finds a new partner and quietly waits for the next prompt from the teacher.
7. After hearing the prompt, students take turns sharing their responses with their partner.
8. This process continues until all of the question prompts are answered.
9. After the activity, call on students to share their responses with the class.

As a general rule of thumb, make sure that the prompts are open-ended or have more than one right answer. If not, the students won't each be able to share, because the first will have answered the prompt. For example, instead of asking partners to tell how many electrons oxygen has, change the prompt to "Tell your partner the number of electrons an element has and have the partner name the element." Ideas for prompts are presented in Figure 3.2.

FIGURE 3.2 *Prompt Ideas for Milling to Music*

General Science	*Earth Science*	*Physical Science, Chemistry*	*Life Science, Biology*
Tell your partner one step of the scientific method and why it is important. (Tell another step, and so forth) Tell your partner one rule for lab safety. (Tell another rule, and so forth) Tell your partner your hypothesis for the experiment. Tell your partner the evidence that backs up the claim you made in your conclusion to the experiment.	Tell your partner one fact about metamorphic rocks. (Tell another fact, or tell a fact about igneous rocks, and so forth) Show your partner what happens at one type of plate boundary and name it. (Show the results of another plate boundary, and so forth) Explain one step of the water cycle to your partner. (Explain another step, and so forth)	Using your periodic table, tell your partner an atomic number and have him/her name the element. (Tell the atomic number of other elements, and so forth) Tell your partner how many protons an element has and have him/her name the element. (Tell how many protons another element has, and so forth) Tell your partner the name of a compound with an ionic bond. (Tell another compound with an ionic bond, or tell a compound with a covalent bond, and so forth)	Tell your partner one example of a mammal. (Tell another example, and so forth) Tell your partner the definition for one vocabulary term in our heredity unit. (Tell another definition, and so forth)

In some cases, the Milling to Music prompt may be a question with one answer if the language objective of the activity is associated with important language functions, such as teaching students ways to agree or disagree, or to paraphrase. In the first case, students are assigned an *A* role or a *B* role. *A* always answers the teacher's prompt and *B* always agrees or disagrees and tells why. In the second case, either student can respond to the prompt first (for example, giving the definition of a key vocabulary word), but the second student paraphrases, or restates the definition in his or her own words.

Biology Science Application

Lesson Concept: Heredity

Content Objective: Students will be able to define key terms in a heredity unit.

Language Objective: Students will be able to confirm, correct, or extend their partners' definitions.

Key Vocabulary: gene, trait, genotype, phenotype, heredity, allele, dominant, recessive, incomplete dominance, heterozygous, homozygous

In a high school Biology classroom, students have recently learned vocabulary for a unit on heredity. The teacher instructs students to take their notebooks with them as they participate in Milling to Music. He plays music and students mill around the classroom. When he stops the music, he prompts the partners to share the definitions for two words, one for each partner. (In this case, it is *phenotype* and *genotype*.) The students are instructed to say the vocabulary word before they say the definition. They confirm or correct their partners' definitions. The teacher then continues the milling activity until all the key words have been defined.

Differentiation

To help students formulate definitions, sentence starters such as the following can be posted and discussed in advance.

Heredity is _____ .
Phenotype means _____ .
Genotype means _____ .

For other types of prompts, different sentence starters may be shared, such as

One fact about _____ is _____ .
One example of a _____ is _____ .

If partners share a native language, and one or both are newcomers or beginners, they may respond in their native language. If one is more advanced, he or she can help the newcomer/beginner articulate the definition in English after it is spoken aloud in the native language.

Practice & Application

- Great Performances
- Musical Flash Cards
- Red Light, Green Light

The SIOP® Model encourages hands-on activities to practice and apply new knowledge using oral and written modes. The following techniques engage students in practicing and applying what they are learning in terms of content and scientific language.

Great Performances (Adapted from *99 Ideas*, p. 120)

SIOP® COMPONENT: Practice & Application

Grade Levels: 4–12
Grouping Configurations: Small groups, whole class
Approximate Time Involved: 5–10 minutes
Materials: Resources for research (e.g., books, articles, websites)

Description

In this activity, students act out or pantomime features of a topic being studied. Students work in groups of three or four to conduct research and act out key information learned. If there are more students than roles, students can share a role. If there are more roles than students, a student can take on more than one role.

When students have practiced their roles, the teacher then puts two or three of the groups together. The original groups take turns acting out their roles for the others in the new larger group to watch. When students have shared their performances in these groups, the teacher may decide to have some or all groups act out their roles for the whole class.

Wrap up the activity by having students write about what they learned.

Grade 3 Science Application

Lesson Concept: Plant Structures

Content Objective: Students will be able to act out the functions of the following plant structures: roots, stems, leaves, and flowers.

Language Objective: Students will be able to summarize the functions of plant structures by completing a cloze paragraph.

Key Vocabulary: roots, nutrients, stems, support, leaves, synthesis, flowers, pollinators, seeds, reproduction

In this lesson, the Great Performances technique reinforces learning that has taken place. The teacher has already taught students about the functions of roots, stems, leaves, and flowers, and students have done some reading about plant structures. Students need not conduct research for this lesson, but will refer back to handouts, charts, and vocabulary with graphics as they work in small groups to apply their knowledge.

(continued)

The teacher first divides students into groups of four. Once in groups, students are each assigned a role of root, stem, leaves, or flowers. As a group, the students decide how they will act out their roles. For example, the student playing the "roots" might sit on a chair and reach downward with their arms.

When students have practiced their roles, the teacher then puts two or three of the groups together. The original groups take turns acting out their roles for the others in the new larger group to watch. When students have shared their performances in these groups, the teacher chooses some groups act out their roles for the whole class.

After students have performed their scenarios and seen other groups perform, they are ready to write a written summary in a cloze paragraph. Students complete the cloze paragraphs individually or with a partner. The cloze paragraph might look like the following:

There are ______________ different parts of a plant. The different parts include the ______________, ______________, ______________, and ______________. Each part of the plant has a different function. The main function of the roots is to ______________. The ______________ provide support. The leaves ______________ while the flowers attract ______________ and produce ______________ for reproduction. My favorite part of the plant is ______________ because ______________. The most interesting thing that a plant does is ______________ because ______________.

Differentiation

More advanced students may write a summary without the cloze paragraph.

Musical Flash Cards

COMPONENT: Practice & Application (also suitable for Interaction)

Grade Levels: 6–12
Grouping Configurations: Partners
Approximate Time Involved: 10–15 minutes
Materials: Flash cards, music and speakers

Description

Musical Flash Cards is a kinesthetic activity that involves all of the students in a class and provides them with an opportunity to practice key content concepts and develop oral language. This activity provides students with a safe environment to practice these concepts and become an "expert" on each question. Musical Flash Cards is a variation of Milling to Music.

The teacher selects practice problems or vocabulary words and makes the flash cards. The flash cards can be created in a word processing program and then printed. To make it simple, the teacher can print the front and back of the card right next to each other. The students can then fold the cards in half before participating in the activity.

Procedure:

1. Each student gets a flash card with a question on one side and the answer on the other.
2. The teacher turns on music.
3. While the music is playing, the students walk around the room.
4. When the music stops, each student finds a nearby partner.
5. Student *A* holds her/his card and reads the question to Student *B*, who may look at the question side of the card and read along. Student *B* states an answer, and Student *A* informs Student *B* if the answer is correct. If not, Student *A* provides correction.
6. Student *B* then holds her/his card and reads the question to Student *A*, who may look at the question side of the card and read along. Student *A* states the answer, and Student *B* informs Student *A* if the answer is correct, providing correction if needed.
7. Student *A* and Student *B* trade cards.
8. The teacher turns on the music again. Students walk around the room until the music stops and then find a new partner.
9. The students continue the process until the teacher tells them to return to their seats.

Musical Flash Cards help students practice and apply what they have learned. Musical Flash Cards also give students a chance to learn from and teach their peers. Moreover, after they trade cards, students become "experts" on their new flash cards. They already know the answer and then can help their next partner answer the question.

Ideas for Musical Flash Card topics are listed in Figure 3.3.

FIGURE 3.3 *Topic Ideas for Musical Flash Cards*

General Science	*Earth Science*	*Physical Science*	*Life Science*
Put a sentence on each card and have the students decide if it is an observation or an inference (middle school). Put vocabulary words on one side and the definition on the other (middle or high school).	Put the name of a rock on one side and have the students determine if it is metamorphic, igneous, or sedimentary (middle school).	Put the name of a substance on one side and have students decide if it is an element, a compound, or a mixture (middle school). Put a physics formula on one side and the formula's use on the other (high school).	Put images of the phases of mitosis on one side and the name of the phase on the other (high school).

Chemistry Application

Lesson Concept: Periodic Table of the Elements

Content Objective: Students will be able to calculate the number of protons, electrons, and neutrons in an element, given its atomic number and atomic mass.

Language Objective: Students will be able to listen to a question and tell a partner the number of protons, electrons, or neutrons in a given element.

Key Vocabulary: element, atomic weight, atomic mass, atomic number, electron, neutron, proton, chemical symbol, periodic table

The students are studying the periodic table of the elements. The students have already learned about atomic numbers, atomic mass, element names, element symbols, protons, electrons, and neutrons. They still need additional practice to master the material because they will need to know how to use the periodic table for many activities in Chemistry class.

The teacher distributes the flash cards he made (see Figure 3.4) and plays the music. The students engage with one another using the procedures listed above. This activity is a great resource to help the students develop oral language, They read one question aloud and then respond to another question aloud. They negotiate meaning as they interact and apply their knowledge of the periodic table. After this activity, they have a better understanding of both the science content of the lesson and the science language involved in the lesson.

Differentiation

Newcomer students may partner with more proficient students and interact as a threesome or as a foursome. Newcomers may be asked to echo read or echo respond once the partner has read or responded.

FIGURE 3.4 *Sample Flash Cards*

Question Side:	*Answer Side:*
How many protons does lithium have? 3 Li Lithium 6.941	3 Li Lithium 6.941 Lithium has 3 protons.
What is the atomic number of silver? 47 Ag Silver 107.87	47 Ag Silver 107.87 The atomic number of silver is 47.

Red Light/Green Light (Adapted from Kagan, 1994)

COMPONENT: Strategies (also suitable for Review & Assessment or Strategies)

Grade Levels: 3–5
Grouping Configurations: Small groups
Approximate Time Involved: 20–30 minutes
Materials: Expository text, chart paper, markers, red tags, teacher-prepared graphic organizers or blank outlines

Description

Red Light/Green Light is an instructional technique designed to help students master and apply content information while learning from one another. For this activity, tables will need to be arranged around the room so that students can work in their groups to complete a task and then some students will rotate to see what other groups have completed. The six steps to the "Red Light/Green Light" instructional technique follow:

1. ***Assign Groups.*** The teacher assigns students to groups of three or four. The teacher may consider language and/or reading ability as she assigns the groups. The goal is to have students grouped heterogeneously. When the groups are created, each group member is assigned a letter. (At this point the teacher may consider assigning A to the most proficient students, B to the least proficient students, and C and D to the students that are at the speech emergent/intermediate fluency levels.) Once groups are assigned, the teacher will assign each group a topic for a poster.
2. ***Groups Create a Poster.*** Once students are in their groups and have an assigned topic, the teacher gives specific directions as to what needs to be included on the poster. The teacher sets a time limit so that all groups finish near the same time.
3. ***Note-taking Preparation.*** Before having groups rotate, the teacher sets the expectations for note-taking at each table station. The students should use a note-taking format that they are familiar with. The teacher might provide a graphic organizer or a blank outline for students to use as they take notes.
4. ***Red Light/Green Light.*** When groups finish their posters, the teacher determines which students get the "Red Light." (If students are grouped heterogeneously, the teacher might have the "B's" receive the "Red Light" first so that the less proficient students are presenting the posters that they created.) Students who receive the "Red Light" (such as a red tag, red piece of paper, etc.) stay at their station to present the poster that they created. The other group members are given the "Green Light" and rotate to the next station. The one group member who stays behind teaches the group that arrives about their poster. Students with the "Green Light" listen to the presentation at their new station and take notes on the information being presented. The teacher should remind students that they have to pay close attention and ask clarifying questions, because they don't know who will have stay and present next. The teacher should also remind students to use their blank outlines, graphic organizers, or other notes to help remember key information to share with the next group.

5. ***Repeat Step 4.*** The teacher now calls another letter to receive the "Red Light" and those students stay to teach the next group about the poster they just learned about. The Green Lighters move on to the next station. The process continues until each student has visited at least three posters/stations.
6. ***Summary.*** Students meet back with their original groups to compare the notes they have taken at each station. The teacher may provide a graphic organizer or a worksheet for students to fill in to summarize their learning.

Grade 4 Science Application

Lesson Concept: Biomes of the World

Content Objective: Students will be able to create a visual representation to show features of an assigned biome.

Language Objective: Students will be able to listen for facts about different biomes.

Key Vocabulary: biome, ecosystem, location, climate
Additional vocabulary is determined by the particular biome a group is studying (e.g., shoreline, sandy, shores, barrier islands, estuaries, salt marshes, tides, waves, currents).

In this lesson, students have already studied 8 biomes (Tropical Rainforest, Tropical Savanna, Desert, Chaparral, Grassland, Temperate Deciduous Forest, Temperate Boreal Forest, and Arctic and Alpine Tundra) and are applying key information that they have learned.

The teacher divides students into small groups and tells them they will need to identify and explain at least five important features about their assigned biome. These five features include the location, information about the ecosystems, climate, and additional interesting facts about the biome. Students are required to create a poster using key vocabulary words and graphics that will help them teach another group about their biome. As students work, remind them that they all need to be able to explain the information on the poster because they don't know who will be required to present the information to the next group.

The teacher sets a time limit for students to complete their posters and prepare themselves to present the information. When time is up, the teacher brings the class back together as a whole group and reviews a blank outline (see Figure 3.5) she has prepared for the students to use for note-taking. She models writing information on the blank outline and then moves students to the next step, identifying which students will stay at their stations. The teacher then asks the other students to rotate clockwise to the next station. When students have rotated, the teacher reminds students that the language objective is to listen for facts about different biomes. The teacher prompts the students that stayed with their poster to begin their presentations. After 5 minutes, the teacher asks students to finish their presentation and prompts students to complete their blank outlines and ask clarifying questions.

The teacher then selects different students to stay and continues to repeat the rotation, presentation, and note-taking steps. When students have visited at least two other stations, the teacher asks all students to return to their original groups.

(continued)

Grade 4 Science Application *(continued)*

When back with their original groups, students are given a blank graphic organizer or outline copied on 11″ × 17″ paper. The teacher models the following sentence starters for students to use in their group discussions:

The ecosystem includes . . .
________ is located . . .
One thing I learned is . . .
The _______ biome has . . .

Each group member uses the sentence starters to share at least one thing he or she learned at each station and students negotiate which pieces of information should be written on the larger blank outline. The students take turns writing on the outline; they can write something from their own notes or something that someone else had written.

Differentiation

Beginning/Early Intermediate: Students at the beginning/early intermediate stages may be partnered up with a more proficient English speaker. If a newcomer, beginning, or early intermediate student is called on to stay and present a poster, the partner stays with that student to support him or her in presenting. This does not mean that the newcomer/beginner/early intermediate student is excused from presenting, but rather that he/she may share the responsibility with a partner. The partner may assist the less proficient student with words and phrases to use.

The teacher may provide sentence starters for students to use in note-taking and in presenting.

FIGURE 3.5 ***Biomes of the World Outline***

Biome:______________

Location:____

____

Climate:____

Ecosystems:____

____

____

____

Interesting Facts:

____

____

____

____

Lesson Delivery

- Nine Squares
- Stop That Video

The techniques and activities for this component should be closely aligned to a lesson's content and language objectives. When teachers incorporate these ideas, they help the students master the content concepts and academic language goals.

Nine Squares

COMPONENT: Lesson Delivery

Grade Levels: 2–8
Grouping Configurations: Individual
Approximate Time Involved: 10–15 minutes, depending on grade level
Materials: Text, blank paper

Description

The Nine Squares is designed to help students learn the cognitive strategy of note-taking while identifying important information about content being presented. As with any technique, the Nine Squares should be modeled and practiced before students apply it independently. The steps for the Nine Squares are:

1. ***Pre-read the text.*** To ensure that the essential information is noted, the teacher should pre-read the text to identify the main points and mark no more than nine places to pause.
2. ***Create the Nine Squares.*** Each student receives a piece of blank paper, preferably without lines. The students fold the paper into thirds vertically and horizontally, creating nine boxes in which to write information. See Figure 3.6.

FIGURE 3.6 *Nine Squares Chart*

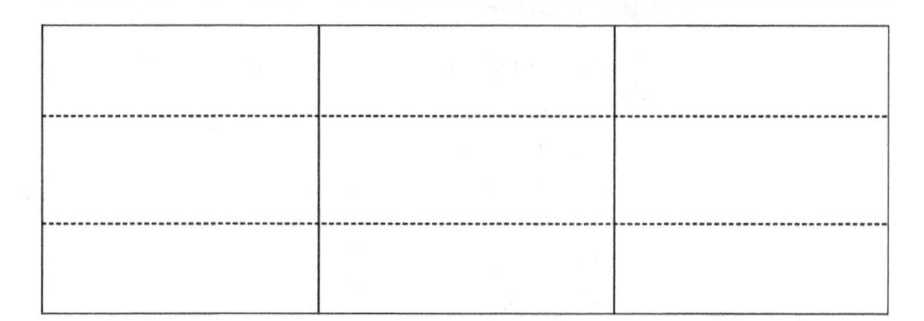

3. ***Read the Text.*** The teacher begins to read the text aloud, pausing when a key point of information is presented. At this moment, the teacher asks the students to discuss the information with a partner and then write 1–3 key words about the point in the first box on their Nine Squares. The teacher continues reading, pausing, and asking students to talk about and note information each time a new, important point is mentioned.
4. ***Summarize the Learning.*** When the Nine Squares is completed, students then summarize their learning by writing two or three sentences on the back of their Nine Squares, describing what they learned.

Grade 2 Science Application

Lesson Concept: Technology and Inventions

Content Objective: Students will be able to explain how the first airplane was invented and built.

Language Objective: Students will be able to take notes on an invention using key vocabulary.

Key Vocabulary: invent, invention, glider, wind tunnel, propeller, engine, wing, wing-span, lift

The class has been studying important technological contributions made by famous people. In this lesson, students are learning about the Wright Brothers and the airplane. The teacher draws a 3 × 3 large square on the board with nine boxes. He then distributes a blank piece of paper to each student. He models folding the paper into nine boxes and circulates the classroom as students fold their own papers.

Next, the teacher reads from a picture book about the Wright Brothers and the first airplane. When he gets to the first key point, he pauses and writes down the first key piece of information on his sample Nine Squares on the board. Students copy the teacher's notes.

The teacher continues to read and stops after the next key point. This time he asks students to discuss what the notes should be. Several students share their ideas aloud and he writes on his sample Nine Squares. The teacher continues this process, pausing after each consecutive key point and allowing students to process and discuss, and then write down their own new information. Periodically, he randomly calls on two or three students to share what they wrote with the whole class. By having students share, the teacher ensure that misconceptions are clarified and that students are keeping up with the pace of the lesson.

When the teacher has finished reading the text and students have completed their Nine Squares, the teacher asks students to summarize their learning by writing two or three sentences on the back of their Nine Squares.

Differentiation

For classes of newcomers or beginners, the teacher may modify the number of boxes in the Nine Squares to four or six squares.

For students at lower levels of English proficiency, the teacher may want to provide the following sentence starters for students to use in their summaries:

I learned . . .
I noticed . . .
I wonder . . .
The most important thing I learned was . . .
Key facts are . . .

For more proficient readers or students who are very familiar with the Nine Squares, the teacher may have students read independently and select their own key information.

Stop That Video (Adapted from *99 Ideas*, p. 174)

SIOP® **COMPONENT:** Lesson Delivery

Grade Levels: All
Grouping Configurations: Whole class
Approximate Time Involved: 15–60 minutes (will differ depending on the video clip and grade level)
Materials: Video or DVD clip, note-taking sheet, graphic organizer, or blank outline

Using a video or DVD can be an effective way to make content comprehensible for learners. However, when videos present a lot of information that students are required to process, confusion may occur. We often hear students talking when educational videos or DVDs are showing; rather than attempting to quiet them, the teacher can provide for structured interactions that help students process the information being presented.

The teacher begins the video and pauses it when key information is presented. At this point, the teacher instructs students to use the key vocabulary as they talk with a partner and note what they just learned from the video. To informally assess comprehension, the teacher may randomly call on students to share the information they discussed with the class before restarting the video. If there are misconceptions, this is the teacher's opportunity to correct or clarify either by presenting the information in a new way or by rewinding the video/DVD clip and working students through it more slowly.

When the video finishes, the teacher can facilitate a wrap-up by having students discuss or write about the important information they learned.

Grade 3 Science Application

Lesson Concept: Light on Objects

Content Objective: Students will be able to categorize different objects based on how light behaves when it strikes them.

Language Objective: Students will be able to use comparative language to describe how light behaves when it strikes different objects.

Key Vocabulary: transparent, translucent, opaque, reflection, absorption

For this lesson, the teacher shows a brief video clip that demonstrates how light behaves when it strikes different objects that are transparent, translucent, and opaque. Before beginning the video, the teacher reviews the key vocabulary words that students have learned in a previous lesson and comparative language phrases such as "Both A and B ___." "C can ___ but D can't ___."

Next, the teacher hands out a blank outline for students to use for note-taking and then begins the video. When the video shows the first example of how light behaves when hitting a certain object, she stops the video and asks students to share with their partners

Grade 3 Science Application *(continued)*

what they just noticed. The teacher then randomly calls on students to share their observations and uses one of the observations to take notes on a model note-taking chart displayed on a transparency (see Figure 3.7).

The teacher continues the video clip and repeats the process. The second or third time she pauses the video, the students talk to their partners and take notes without a teacher model. The teacher asks students to nominate a classmate to share out each time in order to assess comprehension and clarify any misconceptions.

When the video has finished, the teacher demonstrates light's behavior with a flashlight to apply what they are learning. She shines the light on several familiar objects (e.g., a mirror, a drinking glass, waxed paper, a book, aluminum foil, white construction paper, black construction paper) and asks students to categorize them as translucent, transparent, or opaque. She also asks them if the objects reflect or absorb light.

The teacher next asks students to use comparative language frames and key vocabulary to describe how light behaves when it strikes different objects. She posts the frames (see below) and has students first write complete sentences at the bottom of their blank outline and then read their responses to a partner.

When light hits _______, it _______, but when it hits __________, it _________.
When light hits both ___________ and ___________, it _________.

FIGURE 3.7 *Light Behavior*

Objects	*Reflection*	*Absorption*	*Details*
Transparent			
Translucent			
Opaque			

Review & Assessment

- Web of Information
- Snowball!
- Pass the Note Card
- Simultaneous Round Table

Activities and techniques for this final component of the SIOP® Model serve several purposes. They review the content concepts and vocabulary goals of the lesson. They help the teacher assess student comprehension of the material presented and they allow the teacher to give feedback to the students on their production of the language and the tasks they do.

Web of Information

COMPONENT: Review & Assessment

Grade Levels: All
Grouping Configurations: Small groups, whole class
Approximate Time Involved: 5–15 minutes
Material: Yarn

Description

Web of Information is another quick way to review content information while allowing students multiple exposures to vocabulary words. With the Web of Information technique, students can practice listening and speaking skills.

To construct the Web of Information, students are first instructed to think of an outcome sentence and use one vocabulary word to explain something they learned about the content they are studying. After this, the teacher then arranges the students to sit or stand in one large circle and shows them a ball of yarn. The teacher holds the ball of yarn and shares an outcome statement. The teacher then holds one end of the yarn and tosses the ball to a student. That student then shares, holds onto one part on the yarn, and tosses the ball to another student. The activity is complete when all students have shared their responses. As the yarn goes back and forth across the circle, it creates a "Web of Information."

Grade 1 Science Application

Lesson Concept: Natural Resources

Content Objective: Students will be able to review the features of a natural resource.

Language Objective: Students will be able to orally summarize their knowledge of natural resources using outcome sentence starters:

(continued)

Grade 1 Science Application *(continued)*

I wonder . . .
I discovered . . .
I still want to know . . .
I learned . . .
I still don't understand . . .
I still have a question about . . .
I will ask a friend about . . .

Key Vocabulary: air, water, soil, trees, forest, wildlife

To wrap up a lesson on natural resources, the teacher directs the students to sit in a circle at the front of the room. The teacher refers students to the posted vocabulary words and posted sentence starters and asks students to think of a response that they will share with the whole group. The teacher asks students to think quietly for 30 seconds before turning to a partner to practice their response. After practicing their responses with a partner, students are ready to share and the teacher begins by tossing the yarn to the first student. The process continues until each student is holding a strand of the yarn and has shared his or her response.

Differentiation

For classes with newcomers or beginners, the teacher may have these students sit with a partner. Before sharing aloud, beginners can practice again with partners. Also, the partner may speak first and then the beginner echoes what the partner says.

Snowball!

COMPONENT: Review & Assessment (also suitable for Building Background)

Grade Levels: 3–12
Grouping Configurations: Individual, partners, whole class
Approximate Time Involved: 10–15 minutes
Materials: Snowball! graphic organizer

Description

In Snowball! students have permission to do something in the classroom they usually get into trouble for doing . . . throw stuff! Snowball! is an interactive writing activity in which students share information with each other.

The Snowball! technique can be used for reviewing facts. In this case, the first student writes a fact and the other two students have to each write a different fact. Snowball! is also a technique for written debate. In this case, the first student makes a claim with evidence and the other two students each respond about whether they agree or disagree with the topic and give evidence for their response in writing.

Before conducting Snowball! for the first time, review the guidelines with students. It is a good idea to have the procedures with graphics posted on the board for students to see. This way, the students will have a clear understanding of exactly what is expected during the activity. It is also beneficial to model for students where they will write their responses on the graphic organizer.

In general, students become so excited about this activity that they follow the directions without any problems!

Procedures:

1. Students get a Snowball! graphic organizer that has three snowflake bullet points (see Figure 3.8).
2. Students listen to the teacher to find out the prompt for Snowball!
3. Students write their response to the prompt next to the first snowflake (see sample snowball in Figure 3.9).
4. Students crumple up their paper into a "snowball."
5. Students wait for the teacher's cue to throw the snowball to a specific location in the room (usually a location in the middle of the classroom) and then they toss their snowballs.
6. When instructed by the teacher, students get out of their seats by table to retrieve a snowball.
7. Students return to their seats, open the snowball, and read the response.
8. Students write their name next to the second snowflake and write a new sentence on the paper. The sentence can be a response to the first student or a new idea.
9. Repeat steps 4–8, this time using the third snowflake.
10. In the last round, students make eye contact with the original owner and then toss the snowball to the original owner (or the teacher may have the students hand the snowballs back).
11. The original owners read the other responses that have been written on their snowball.

FIGURE 3.8 *Snowball! Graphic Organizer*

Name_______________________

Snowball!

____
Name 2

____
Name 3

FIGURE 3.9 *Procedures with Graphics for Snowball!*

Procedures for Snowball

1. Put your name on your paper.
2. Listen for directions from the teacher.
3. After you hear the directions, write your first sentence.
4. Crumple up your paper like a snowball!
5. When the teacher says, throw the snowball to the designated area.
6. When the teacher tells you to, go get a snowball. Read it. Add a second sentence.
7. Repeat steps 4–6 and add a third sentence.
8. When the teacher tells you to, crumple the snowball and toss it to the original owner.
9. Read your original snowball and be prepared to share.

Some ideas for Snowball! prompts can be found in Figure 3.10.

FIGURE 3.10 *Sample Prompts for Snowball!*

General Science	*Earth Science*	*Physical Science*	*Life Science*
One important lab safety rule is . . .	One fact I learned about volcanoes is . . .	My favorite element is ___ because ____.	Genetic engineering is ethical/non-ethical because ____. *
My hypothesis is . . .	If a tornado is approaching, one thing I should do is . . .	The most important mineral is ____ because ____.	My favorite sense is ____ because ____.
My favorite scientist is____ because ____.	The worst kind of pollution is ____ because ______.*	The United States should/should not have nuclear weapons because ______.*	

* With these prompts, students may respond to first student with "I agree/disagree with _____ because _____."

Biology Application

Lesson Concept: Genetic Engineering

Content Objective: Students will be able to use evidence to make a claim about the ethical nature of genetic engineering.

(continued)

Language Objective: Students will be able to respond to another student's opinion by writing whether they agree with that student.

Key Vocabulary: genetic engineering, opinion, evidence, claim, traits, ethics, ethical, nonethical, support

In a high school Biology class, students are learning about the pros and cons of genetic engineering. The teacher wants students to form an opinion about whether they think genetic engineering is ethical and wants students to use evidence to support their answers. The teacher also wants to create discourse among the students.

The teacher gives students the following prompt for Snowball!: "Genetic engineering is ethical/nonethical because _____." Students write their responses on their snowball organizers and toss them. The second student, who retrieves a snowball, agrees or disagrees with the first student and this creates a written dialogue between the students in which they are practicing and applying their language. The second student tosses the snowball and the third student who retrieves it, writes his/her response after reading the two already on the page.

When the third and final round of Snowball! is complete, the teacher randomly calls on students to share their original claim and evidence. After each student reads her/his claim, the teacher will encourage other students to agree or disagree.

Differentiation

If the class has newcomers or beginners, the teacher may pair the lower level students with the advanced students so they work as a team on one snowball.

The Snowball! technique can also be used as a building background activity for students to share what they know about a topic before a unit begins or to recall information from a prior lesson.

Snowball! may also be modified as an oral interaction. After students write a response to the first prompt, they are divided into two groups. Group 1 tosses the snowballs and Group 2 retrieves and finds the author. They discuss their responses. Then Group 2 tosses their snowballs and Group 1 retrieves, ideally collecting one from a different partner. These new pairs discuss.

Pass the Note Card

COMPONENT: Review & Assessment

Grade Levels: 3–12
Grouping Configurations: Whole class
Approximate Time Involved: 5–10 minutes
Materials: Index cards

Description

Pass the Note Card is a quick way to review content information while allowing students multiple exposures to vocabulary words. With Pass the Note Card, students can practice the four domains of language: reading, writing, listening, and speaking.

The steps for Pass the Note Card are simple. The teacher distributes a blank index card to each student. Students are instructed to use at least one sentence starter and one vocabulary word to write something they learned on the index card. The teacher then has students stand in a large circle, each with an index card in his/her hand. The teacher plays music. While the music plays, students pass the index cards, one at a time. When the music stops, the students read what is written on the index card they have in their hand. The teacher can then randomly call on students to share the information on the card with the whole circle. When the music starts again, the steps repeat.

Grade 4 Science Application

Lesson Concept: Vertebrates

Content Objective: Students will be able to identify facts in the text about vertebrates.

Language Objective: Students will be able to use key vocabulary and sentence frames to summarize important information about vertebrates.

Key Vocabulary: mammals, birds, fish, reptiles, amphibians

At the end of a lesson during which students studied the characteristics of vertebrates and different types of animals that are vertebrates, the teacher passes out blank index cards and refers students to the vertebrate word wall and the posted sentence starters:

Vertebrates have . . .
Some vertebrates are . . .
One interesting thing I learned about vertebrates is . . .
I would like to know more about . . .
Vertebrates are/are not interesting because . . .

The teacher asks students to write one sentence on each side of the card to summarize facts about vertebrates. On one side they can choose any sentence starter to use. On the other side, students must use the last sentence starter (Vertebrates are/are not interesting because . . .). Students are required to use the last sentence starter to ensure that they are using some higher order thinking skills.

When students have their responses written, they form a large circle in the front of the room. The teacher plays music, and students pass their index cards, one at a time, to the right. When the music stops, the teacher asks students to silently read the information on the card. Next, each student turns to a partner and whisper-reads the information on the card aloud. The teacher then calls on a few students with a particular characteristic (such as someone wearing white shoes) to read the information on the index card in their hand to the whole class. The class repeats the activity five to seven times, with the teacher calling out different characteristics for identifying students to read aloud (such as students born in February).

(continued)

Differentiation

The teacher may use a buddy system with one student who can write and read paired with one who cannot. The buddies discuss the facts and choose a sentence starter. The first partner writes on the index card. If called on once the music stops, the first one reads and the second one repeats, if possible.

Some students might draw pictures to show what they learned.

For students who are familiar with the process, the class may be divided into two or three smaller circles. The teacher moves from circle to circle as students share information.

Simultaneous Round Table (Adapted from *99 Ideas*, p. 178)

COMPONENT: Review & Assessment

Grade Levels: 4–12
Grouping Configurations: Small groups
Approximate Time Involved: 5–15 minutes
Materials: Paper and pencil

Description

Simultaneous Round Table is a process for students to quickly generate responses to a prompt. Students pass papers around the table and record their responses. Not only do they write their own response, but they also read and learn from the responses of others at their table. This technique works especially well to review information at the end of a lesson or a unit.

Procedures:

1. Each student gets out a pencil and a piece of paper.
2. The students write the names of all students at their table on the paper.
3. Students listen to the teacher for the writing prompt. A prompt should be designed so that it offers many possible responses. (See sample prompts in Figure 3.11.)
4. Each student writes a response to the prompt.
5. When every student at the table is done with his/her response, the students pass their papers clockwise to the next person.
6. Students read the answers on the paper and add a different answer to the list. They can add a new idea or they can write an idea they read on another paper.
7. The students continue generating responses to the prompt and passing the paper around the table until the teacher tells them to stop.
8. At the end of the activity, the students share their ideas aloud at the table and then the teacher calls on students to share with the class.

FIGURE 3.11 *Sample Prompts for Simultaneous Round Table*

General Science	*Earth Science*	*Physical Science*	*Life Science*
Make a list of tools you can use to measure things (middle school) List famous scientists and their contributions (high school) Write observations you made during the experiment (or field trip)	List the famous volcanoes (middle school) List celestial objects (middle school) List factors that contribute to air pollution List types of energy and give examples (middle school)	List as many elements as you can (middle school) List as many minerals as you can (middle school) List examples of forces in your daily life (middle school) Give examples of potential energy being converted to kinetic energy (high school)	List the parts of a microscope (middle school) List the parts of plant/animal cells (high school) List the bones/muscles/tendons in the body (high school) List genetic disorders (high school) List factors of a specific biome

Physics Application

Lesson Concept: Transformation of Energy

Content Objective: Students will be able to provide examples of energy transformations in their daily lives.

Language Objective: Students will be able to record and read a list of energy transformations.

Key Vocabulary: transformation, conversion, potential energy, kinetic energy, electrical energy, mechanical energy, chemical energy, thermal energy, light energy, solar energy, sound energy

In a high school Physics class, the students have already learned the basics about energy transformations and have learned about some specific examples of these transformations. The teacher wants the students to review energy conversions so she can assess their learning. The teacher decides to use Simultaneous Round Table so all of the students will be writing, reading, and learning from others in their group.

The prompt for Simultaneous Round Table is "List as many examples of energy transformations in your daily life as you can." Because this activity is very specific, the teacher asks students for input in order to model a few examples before the students start. Some of the energy conversions that they might generate are:

> Potential energy is converted to kinetic energy when you drop a basketball from your hands.
> Electric energy is converted to sound energy when you turn on your radio.

(continued)

Mechanical energy is converted to heat energy when you rub your hands together.
Solar energy is converted to electrical energy in a solar panel.

The students benefit from this review activity because they record their own ideas and also read the ideas of others in their group. By the end of the activity, they have a much longer list than what they would have generated on their own.

Concluding Thoughts

In this chapter, we have presented a variety of techniques and activities for the final four components of the SIOP® Model and applied them to classroom scenarios. It is important to keep in mind that although we described these techniques by SIOP® component for organizational purposes, the techniques are intended to support each lesson's language and content objectives; they should not be viewed or used as isolated activities. Our research has shown that lessons are most effective for English learners when all of the components of the SIOP® Model are incorporated. Lessons are ineffective when teachers pick and choose only their favorite SIOP® features. Go ahead and try out these techniques in your lessons. We hope your students will find them as meaningful and engaging as ours have!

Revisiting Mr. Etherton

Douglas Etherton (who was introduced in Chapter 2) has continued to work with his SIOP® coach and other teachers at his school to implement the SIOP® Model. Over the past month, he decided not to give up on science experiments, but has spent time planning to incorporate techniques that support students in conducting the experiments. Using ideas he observed Mrs. Graham use in her model lesson, Mr. Etherton now includes written procedures with graphics for every experiment that students are expected to conduct. He also models the procedures for the experiments as he introduces them. Next, he goes a step further and models exactly how to document notes for the experiments, offering sentence starters and the correct use of vocabulary words. He has eliminated worksheets that may be confusing and has selected worksheets or observation forms that scaffold the data collection and learning. When necessary, Mr. Etherton creates a data collection form. He has continued to ask students to explain the directions back to him, but he now asks students to paraphrase the directions in their own words.

Mr. Etherton also decided to teach students specific roles for conducting experiments (e.g., recorder, reporter, materials manager). He no longer allows the students to determine who will complete which part of the activity; he assigns the roles. He does try to ensure that no matter what role a student is assigned, each person has a responsibility that requires him or her to demonstrate progress toward meeting the academic content and language objectives of the lesson. Further, he notes the roles assigned in his plan book and rotates roles among students. For those with lower proficiency in English, he waits until later in the quarter to assign roles with higher language demands, such as reporter.

When Mrs. Graham asked what he has noticed after making the adaptations, Mr. Etherton responded, "I am definitely spending more time preparing my lessons, but it is paying off. I didn't know where I would even find the time to do this, but I've found a balance. I'm spending less time trying to figure out how to re-teach material that the students didn't understand and I'm even spending less time handling discipline problems. I would much rather devote my time to preparing a successful lesson than writing referrals, calling parents about detentions, and cleaning the room after class. I'll admit, I think I had really lowered my expectations of the students. Now, with my support, they are safely and successfully conducting the experiments. Now that the experiments are running more smoothly, they are also able to correctly complete their data tables and they use the key vocabulary regularly!"

SIOP® Science Lesson Planning and Unit Design

Mrs. Allen's Vignette

Mrs. Evelyn Allen is a ninth grade Earth Science teacher from St. Paul, Minnesota. She teaches science throughout the day, and has some English learners in each class. One class in particular includes 10 native English speakers, and 15 English learners of all proficiency levels. The ELs' native languages include Hmong, Somali, Amharic, and Spanish.

Mrs. Allen is being asked to implement the SIOP® Model in all of her science classes, but she shares some concerns with Mr. Myers, a colleague in the science department. "I've learned the SIOP® Model, and some of it could be helpful, but some features seem to contradict our science methods that include discovery-based learning. I think discovery learning is more important than SIOP® in my science classes, but I am required to use the SIOP® Model. I don't know how to do both. If I post and explain objectives, key

vocabulary, and procedures for the experiments to the students, then the students really won't be discovering anything on their own."

After a pause, she continued, "I guess I do have to change something however, because when I try discovery-based learning, it isn't as successful as I want it to be. When I just give them materials without directions or an explanation, they aren't making academic scientific discoveries. They're really just playing with the materials."

Mr. Myers asked Evelyn how she plans her lessons and units. She explained that she looks at the district curriculum framework and pretty much follows the textbook. "I try to make sure I pay attention to the topics that we know will be on the state science test," she added. Mr. Myers encouraged her to think more broadly about lesson planning. By using the SIOP® Model, he explained, she could think through the academic English needs the students have as well as the science language demands of the lessons and lab experiments. "What do you want your students to learn for each unit you teach? And not just the science facts. What reading, writing, listening, and speaking skills? What vocabulary? How do you want them to think about scientific phenomena? Which reasoning processes do you want them to use, and how can you help them articulate their reasoning? Try starting from this perspective as you design an Earth Science unit."

Taking Mr. Myers's advice, Evelyn made a list of questions to guide her in determining the learning goals, how all students will attain them, and how she would know if the students met the goals. She asked herself:

1. What must students learn in this unit? What are my content and language objectives?
2. How will I assess their learning of the unit's content concepts?
3. Where do I need to start to make sure students meet the learning goals identified in Question 1?
4. Is the time allotted sufficient to ensure that all my students meet the content and language objectives?
5. How many lessons will it take to teach the concepts and lead students to mastery?
6. What materials will I use to supplement the textbook?
7. What vocabulary must they already know and what additional vocabulary must be taught?
8. How will I make the lessons meaningful so students are engaged and motivated to learn?
9. How will the language skills of reading, writing, listening, and speaking be incorporated so all students practice these skills in the context of Earth Science?
10. What additional content concepts should be taught to students who have had their education interrupted or who might lack the English proficiency required to comprehend the content and language objectives?

Evelyn considered these questions and began planning. She identified five major concepts from the state standards and three language functions she wanted students to master. She thought about a project they could do and how she'd assess it. She was on her way!

Introduction

This chapter focuses on unit design and lesson planning processes that science teachers undertake as they implement the SIOP® Model. Its aim is to demonstrate how meaningful, engaging activities with academic language practice, such as those described in Chapters 2 and 3, can be incorporated into science lessons and units. Effective SIOP® teaching includes explicit instruction of the content material and associated academic language, plus teacher modeling, guided practice, and independent practice.

As educators we seek the gradual release of responsibility, so our students can engage with science tasks independently. (See Figure 4.1.) As teachers, we present information and model through focused mini-lessons "the type of thinking required to solve problems, understand directions, comprehend a text, or the like" (Fisher & Frey, 2008, p. 5). We then plan guided instruction activities so we can monitor student practice and application of the information from the mini-lesson. We give students an opportunity to collaborate so they discuss ideas and information they learned during the focus lessons and guided instruction—not new information. As we plan lessons within units, we incorporate techniques and activities for students to apply learning strategies, answer higher order questions, and use hands-on materials and manipulatives. When students have acquired background knowledge from the first three phases and are ready to apply the information, we offer them independent tasks. For English learners, this process may take some time, depending on their academic English proficiency and educational backgrounds.

In our own practice of planning and delivering effective SIOP® instruction, we also consider what the final product, outcome, or academic expectation is for all students based on their literacy and language abilities. The sample lesson plans included in this chapter demonstrate how the objectives drive the selection of meaningful activities and the vocabulary that must be developed; the process the teacher will take to guide students in mastering the concepts and language; and the way students will demonstrate their understanding and progress in meeting the learning goals. You'll see that the units in Chapters 5–8 wrap up with a project or other type of assessment so the teachers can be sure the students have mastered the content and language objectives.

FIGURE 4.1 ***Gradual Release of Responsibility***

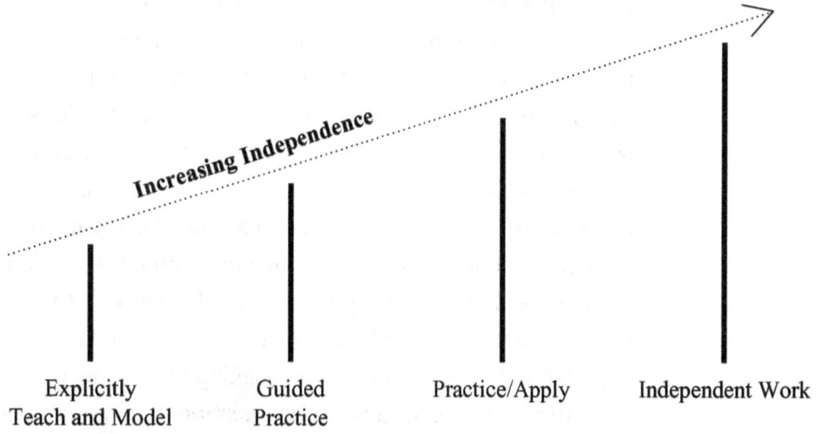

From: Echevarria, J., Vogt, M.E., & Short, D. (2004). *Making Content Comprehensible for English Language Learners: The SIOP® Model*, Second Edition. Boston: Allyn & Bacon.

SIOP® Unit Design

We know from our interactions with teachers, schools, and districts implementing the SIOP® Model that there are multiple ways to design SIOP® lessons and units. Some are prescribed by the school or district; some are individual department or teacher preferences. All models are feasible if the lessons incorporate all the SIOP® features. In some districts like Clifton (NJ) Public Schools and Charlotte-Mecklenburg (NC) Schools, teachers have been paid for summer curriculum work to design new SIOP® units or SIOP-ize existing units. In other sites, such as Lela Alston Elementary School (Phoenix, AZ), grade-level teams have worked together to write language and content objectives to be used in units throughout the school year. Some districts have begun posting units and lessons on the district website.

For those of you who do not yet have an established process for developing your science units, we would like to share the SIOP® unit planner that we have used with several districts during our SIOP® research projects. Figure 4.2 lists the steps we have undertaken with teachers to design the units. In the first step, we identify the critical science topics to cover in a given unit. This is usually done by reviewing the state science standards, the local curriculum, required laboratory experiments, and the textbook. In some cases, experienced teachers with knowledge of the state science tests may also suggest critical topics to include. During this step a large list may be created, but it may be winnowed down as essential topics are highlighted.

The second step derives from the first. Looking at the content topics and the likely materials to be used (chapters, lab directions, etc.), teachers begin listing scientific terms, general academic words, and language functions that may be taught in conjunction with the topics. Teachers may consult the state English language proficiency standards and the English language arts standards. They consider listening, speaking, reading, and writing skills needed to complete tasks in the unit and their students' proficiency levels.

The first two steps may be done on scratch paper. It is in Step 3 where teachers start completing the SIOP® unit planner. The template in Figure 4.3 is used as a brainstorming tool at the start of the process. Teachers first list the key concepts they decided on and then draft the content and language objectives. As they work through the rest of the planner, teachers may return to these sections to adjust the objectives. Usually the objectives are further refined during the actual writing of the lesson plans. Because some schools and districts require teachers to indicate what particular standards are being met in the units or lessons, teachers may want to write reference notes to the state science and language standards on the planner as well. Finally, teachers record related topics that have been studied and might be useful for making connections.

Step 4 is very important for the unit and is the crux of what will make the lessons appropriate for English learners. Here, teachers generate a list of activities they might include in the unit that will advance their students' academic language skills, yet be applicable to the science concepts being taught. These activities will be developed to promote scientific and other academic vocabulary knowledge, reading skills, writing assignments (e.g., lab reports), listening and speaking tasks, and, if feasible, related grammar points (e.g., the formation of a research question, the verb tenses in an if-then statement). In addition, learning strategies that might be useful (e.g., how to use Internet resources, how to take notes on a T-Chart, how to memorize key facts) could be planned. As noted, this is

FIGURE 4.2 *SIOP® Unit Planning Steps*

Step 1: Identify the Content Topics

- Review the state content standards, grade-level curriculum, textbook, tests
- Examine graphs, charts, tables, diagrams
- Decide what's essential

Step 2: Identify Language and Literacy Skills

- Review the state ESL and ELA standards
- Review the grade-level curriculum and textbook to
  - Identify key scientific terms and academic words
  - Determine text structure and language functions
- Consider the reading and writing skills needed for the unit
- Consider tasks students will be assigned and the embedded language of the language tasks

Step 3: Begin the Unit Planner

- List the key concepts
- Draft the content objectives
- Draft the language objectives
- If useful, reference related state content and language standards

Step 4: Generate Language Activities

- Brainstorm activities for vocabulary, reading, writing, grammar, listening/speaking while keeping the content topics in mind
- Identify possible learning strategies students might practice

Step 5: Gather Supplementary Materials

- Secure materials and multimedia that build background
- Look for related fiction or expository readers that have instructional scaffolds in the text, or offer on-level options
- Collect and organize hands-on materials such as manipulatives and realia

Step 6: Plan Lessons

- Organize content into lessons
- Use SIOP® lesson plan template
- Incorporate "interactivities," language practice, and supplementary materials
- Establish a unit project or other unit assessment

Adapted from Short, D. (2009, June). *Sheltered instruction: Curriculum and lesson design.* Paper presented at the 31st Sanibel Leadership Conference, Sanibel, FL. Used with permission.

a brainstorm list. Not all the activities may end up being used in the lessons; some may turn out to be better for later units in the course.

For Step 5, teachers begin gathering the supplementary materials they will need for the unit. For science teachers, first and foremost this means assembling the appropriate supplies for the experiments that will be conducted. Other materials that can build background or make the content more comprehensible (e.g., visuals, multimedia, manipulatives, computer programs) will be beneficial for English learners. Scaffolds such as graphic organizers, study guides, glossaries, and the like could also be collected.

The final step of the unit design process is writing the SIOP® science lessons. This step will involve some adjustments to the unit planner as the content topics are organized into individual lessons and the objectives and tasks are allocated. As the writing takes place, the content and language objectives will be sharpened. Some new tasks may come

FIGURE 4.3 *SIOP® Unit Planner*

Subject: ______________________

Unit Focus: ______________________

Topic: ______________________

Related Topics:

Key Concepts	*Content Objectives*	*Language Objectives*

Unit Project:

Vocabulary Tasks	*Reading Tasks*	*Writing Tasks*	*Speaking/Listening Tasks*	*Grammar Focus*	*Student Learning Strategies*

to mind and replace ones in the planner. Those changes are to be expected. This is also the step where teachers can create a unit assessment, perhaps a project students will do, a performance they might present, a writing piece for their portfolio, or another significant wrap-up activity to measure the knowledge they have gained during the course of the unit.

SIOP® Lesson Planning

When planning effective SIOP® lessons, we always begin with explicit content and language objectives that are derived from content-specific standards and the academic language needs of our English learners. We know from our own experience as SIOP® teachers and SIOP® professional developers that language objectives are critically important to the development of academic language. As you noticed in our application sections of the techniques in Chapters 2 and 3 and as you will see here in the lesson plans, science topics lend themselves to a rich variety of language objectives. This is a good thing! We want students to learn vocabulary, language structures and functions, reading and writing, and listening and speaking skills.

If you use our SIOP® unit planner or a similar tool, you will already have a list of potential content and language objectives to draw from. In addition, you can review the six types of language objectives in the core SIOP® texts (Echevarria, Vogt, & Short, 2008, pp. 29–30; 2010a, pp. 32–33; 2010b, pp. 32–33). It is important to vary the types of objectives our students experience and the tasks they practice them with. A singular focus on only vocabulary or only reading comprehension strategies will not serve English learners well.

Besides writing the objectives in the lesson plan, we need to present them to the students each day. In Figure 4.4, Melissa Castillo, a SIOP® National Faculty member, suggests a number of activities for making content and language objectives a relevant part of daily lessons (Castillo, 2008). We hope you find them useful in your own practice.

If you use one of the lesson plan formats from the other SIOP® books, you know that it includes reminders of SIOP® features that should be present in the lessons. Incorporating all thirty features is a challenge, but well worth it for your students' academic progress. We encourage you to keep the SIOP® Protocol handy as a lesson plan checklist. The more you reflect on how the features can be manifested in your lessons, the more comfortable you will become with the writing process. Over time, including the features in your lessons will become a habit.

You have probably discovered that writing SIOP® science lessons that incorporate the components and features of the model requires careful, detailed planning. However, as shown in Figure 4.5, as you maintain the practice of including all features in your science lessons, you will find that it becomes your internalized way of teaching, and less detailed lesson plans are needed over time.

For further assistance with SIOP® science lesson planning, see the recent brief from the national Center for Research on the Educational Achievement and Teaching of English Language Learners (CREATE), "*Using the SIOP® Model to Improve Middle School Science Instruction*" (Himmel, Short, Richards, & Echevarria, 2009) and the chapter in NSTA's book *Science for English Language Learners* entitled "Designing Lessons: Inquiry Approach to Science Using the SIOP® Model" (Echevarria & Colburn, 2006).

FIGURE 4.4 *Ways to Present Objectives to the Class*

Read the objective as a shared reading piece with your entire class. Then ask students to paraphrase the objective with a partner, each taking a turn, using the frame: We are going to learn ____.

Ask students to read the objectives on the board and add them to their Science Learning Notebooks in a paraphrased form. Then have them read their paraphrased objectives to each other. This could be done as a sponge activity during the first part of the class while you are taking roll!

Present the objective and then do a **Timed Pair-Share**, asking students to predict some of the things they think they will be doing for the lesson that day.

Ask students to do a **Rally Robin** (taking turns), naming things they will be asked to do that day in that particular lesson.

Ask students to pick out important words from the objective and highlight them, such as the action words and nouns.

Give students important words to "watch" and "listen" for during the lesson and call attention to that part of the objective when you mention the academic vocabulary in the lesson.

Reread the objective using shared reading during the lesson to refocus your students.

Ask students to rate themselves on how well they are understanding and meeting the objective, using finger symbols that can be shown in class or hidden under the desk. For example: Thumbs up: I got it! Thumbs down: I am completely lost! Flat hand tilting back and forth horizontally: I understand some of it, but I'm a bit fuzzy!

Rate yourself 1–3, how well did you meet the objective today?

1. I can teach the concept to someone else because . . .
2. I can demonstrate my learning and want to know more . . .
3. I'm not sure, I need more . . .

Ask students to write one or two sentences explaining what they learned in class and give an example. This can be done on an index card, in a learning log, or on a post-it note left on the desk.

Have students do a **Round Robin** (taking turns talking for a specified time with a partner) about how they can prove they met their learning objective for the day.

Use **Tickets-Out**. Students write a note to the teacher (or a letter to a parent) at the end of the lesson telling what they learned and asking any clarifying questions they need answered.

FIGURE 4.5 *SIOP® Lesson Planning Over Time*

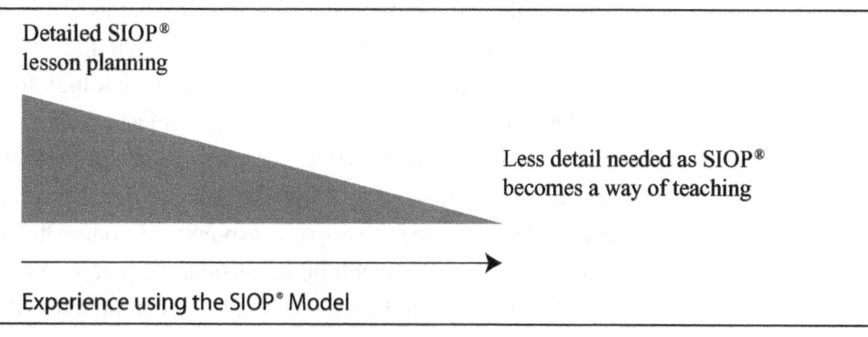

Technology in SIOP® Lessons

The increasing use of technology in the classroom is an exciting trend, filled with many possibilities. Ten years ago, using PowerPoint slides in a lesson was innovative, but now it seems to be a standard and the students themselves are adept at designing animated slides and embedding video clips. More and more teachers use interactive white boards (e.g., smart boards, Promethean boards) as replacements for traditional whiteboards or overhead transparencies. Information can be projected, color-coded, merged, and edited. Some interactive whiteboards allow teachers to record their instruction and post the material for review by students at a later time, providing much-needed opportunities for repetition for English learners.

Multimedia tools now not only allow teachers and students to access information but also to record and present information. Many districts have online video subscriptions (e.g., through discoveryeducation.com or linked to their textbooks) and teachers no longer have to manage TV/video players or (to really date us) filmstrip projectors. With these video subscriptions and access to digital libraries, teachers regularly identify video clips that build background for students, on subjects ranging from volcano eruptions to honeybee behaviors to crystallization of snow flakes. Digital video and audio recording instruments are available with editing capabilities so teachers and students can fine tune a performance or project. Students not only listen to podcasts now, they can create their own. Think about an online science fair where—through video and audio recordings and live web feeds—students around a state can present their projects and judges can ask questions in real time and assess them from afar.

Technology tools for students to use independently are abundant and have moved beyond a computer "spell-check" function. Some interactive software programs such as *Inspiration*, *Kidspiration*, and *Language Learner* are available, as are interactive websites. Computerized reading programs have online coaches that monitor student comprehension of online text and adjust the readings offered as well as the questions posed based on student scores. With these programs, students can also record their own speech and check their pronunciation and fluency. Many of these programs print reports for teachers that show how well students are progressing.

In addition, some classes have access to hand-held scientific equipment (e.g., probeware) that transfers information from experiments right into computer programs. For instance, if students are examining substances for potential chemical reactions, a probe can take the temperature of a substance before and after another substance is added to it and the computer will note any changes, collate the data, and analyze it.

Other teachers we have observed use technology for group-administered reviews. Students hold a device (known as a *clicker*) and respond anonymously to a prompt from a teacher with yes/no, true/false, or multi-choice options. The computer records all student responses and through an LCD, the data are quickly collated and displayed, so the teacher and students can see the correct responses. At once, the teacher can determine if some students may need reteaching of particular concepts.

Technology can be utilized in all the components of the SIOP® Model, as the following examples reveal:

Preparation: multimedia and other technological tools to incorporate in lessons, online lesson planners

Building Background: digital libraries and video clips to build background on scientific phenomena, online visuals for vocabulary development, digital photographs of science experiments to recap activities from prior lessons

Comprehensible Input: software for semantic webs, concept maps, and graphic organizers, audio recordings of textbooks, embedded glossary links for words in online text

Strategies: audio recordings of think-alouds and other problem-solving processes, online note-taking charts and outlines

Interaction: student-developed and edited podcasts and video stories using digital media

Practice & Application: probeware for science experiments, computerized reading and writing programs

Lesson Delivery: web timers to keep teachers and students on pace, multimedia to heighten student interest and engagement

Review & Assessment: hand-held clickers for quick informal assessment of the student learning in class, data management tools, computer adaptive tests and quizzes

By all means, this list is not exhaustive. Its aim is to show that it is not difficult to embed technology in SIOP® lessons if you have access to high-tech and low-tech tools.

However, we recognize that not all schools have these resources available due to lack of funds or a low commitment to technology use. Often, it is the schools with high numbers of English learners that are the most disadvantaged. In order to be sensitive to teachers of ELs who do not have access to state-of-the-art technologies, the lesson plans presented throughout this book are fairly "low tech." If you are one of the fortunate teachers who have these resources at your disposal, when the lesson plan calls for use of an overhead transparency or worksheet, you might use an interactive white board; when the lesson plan suggests showing a picture, you might project a website. We encourage the SIOP® science teachers to integrate technology in their lessons to the extent feasible and appropriate.

SIOP® Science Lesson Format

At present, there are a number of SIOP® lesson plan formats that teachers are using in schools and districts throughout the country. Some have been created by teachers, others have been adapted for the SIOP® Model from district lesson plans, and others have been created by SIOP® National Faculty and the SIOP® authors. You may remember that four lesson plan formats are included in the core text, *Making Content Comprehensible for English Learners: The SIOP® Model* (Echevarria, Vogt, & Short, 2008a), and additional sample lesson plans and formats are included in *Implementing the SIOP® Model through Effective Coaching and Professional Development* (Echevarria, Short, & Vogt, 2008b) and *99 Ideas and Activities for Teaching with the SIOP® Model* (Vogt & Echevarria, 2008). Although the lessons in this book use one format, we, the SIOP® authors, do not endorse any particular lesson plan format. We advocate that you select one that works well for you. You may adjust one of these SIOP® lesson formats to your preferences and district guidelines, and that is fine for SIOP® lessons, as long as all of the components and features are included.

The lesson plan format used in this book was adapted from a template developed by Melissa Castillo and Nicole Teyechea, SIOP® National Faculty members. Hope Austin-Phillips, a practicing science teacher, and Amy Ditton, a former teacher and current SIOP® National Faculty member, are the designers of the science lessons and units.

Figure 4.6 explains the different elements of the format used in this book.

FIGURE 4.6 *Sample SIOP® Science Lesson Plan Template*

Grade(s): ___ **Topic:** *list the topic of the lesson*

Key: SW = Students will **TW** = Teacher will **HOTS** = Higher Order Thinking Skills

Lesson Title: *List lesson title* **Grade(s):** *List grade*

Science Standard: *List state science standard(s)*

Key Vocabulary:	**Visuals/ Resources/ Supplementary Materials:**
List key vocabulary (scientific and academic) **HOTS:** *List higher-order questions or tasks*	*List supplementary resources*

Connections to Prior Knowledge/Provide Background Information:
list ideas and activities for building background or connecting to prior lessons

Objectives:	*Meaningful Activities:*	*Review/Assessment:*
Content Objectives: List the content objectives **Language Objectives:** List the language objectives	List in chronological order, the activities or steps of the lesson	As appropriate, list review and informal/formal assessments next to the corresponding activities

Wrap Up:
list a review or wrap-up activity
remember to review the vocabulary and key concepts
check if the objectives were met

(Template developed by Melissa Castillo and Nicole Teyechea. Used with permission.)

Sample SIOP® Science Lesson Plans

In this section, we illustrate how a full SIOP® science lesson plan can be designed using the techniques and activities we explicated in the previous two chapters. We present four lessons, one per grade-level cluster, on four different science topics. Note that in the plans, there are techniques and activities for students to build background, apply learner strategies, answer higher-order questions, use hands-on materials or manipulatives, interact with the teacher and other students, and review key content concepts and vocabulary. The teacher instruction should facilitate student learning of the content and language objectives, with a variety of techniques to make the content comprehensible and multiple indicators to assess student comprehension. In later chapters, you will revisit these lessons and see how they are positioned in complete SIOP® science units.

It is important to our contributors that each lesson give students the opportunity to build background knowledge of science concepts through hands-on experiences. They must also be introduced to key terms and learning goals. The background building activities provide the students with fresh, concrete knowledge upon which they can build understanding of more abstract concepts and through which they can begin to recognize any misconceptions they may have had.

These lessons also provide numerous opportunities for students to develop academic proficiency in the language of science. Not only are they taught important scientific terms and how to use them in context but they are also taught how to read, write, speak, and listen like scientists. The lessons present and encourage students to use sentence starters and sentence frames in their oral and written discourse. Students learn to make hypotheses, describe their observations, give evidence to justify claims, state their opinion, argue a counter position, articulate conclusions, and more. In other words, students practice accountable academic talk in the science classroom.

We want to point out that all eight components of the SIOP® Model and its thirty features work together to make a lesson effective for English learners. Although we have not identified each SIOP® feature in the following lesson plans, our hope is that you can envision how they have been operationalized and can recognize how they come to life in the delivery of the lesson. Some features will be very clear, such as making connections to prior lessons (feature # 8 of Building Background) and using hands-on manipulatives (feature #20 in Practice & Application). Others require your visualization, such as wait time (feature #18 of the Interaction component), which is not designated in the lesson plan itself, but it is assumed that an effective SIOP® teacher will ensure all students will be provided the time they need to meet the objectives of the lesson. Student engagement (feature # 25 of Lesson Delivery) is also not explicitly stated, but as you read the meaningful activities included in the lessons you can conclude that students will be very engaged.

In the lesson descriptions, you will also notice that there are some teacher Think-Alouds and Planning Points. As experienced SIOP® teachers, Amy and Hope have included their own thought process for planning SIOP® lessons. The teacher Think-Alouds reflect decisions they made or questions they asked themselves during the planning process to lead to high-quality teaching. The Planning Points offer some specific information about useful resources.

The lessons are designed to show how SIOP® features are integrated into science lessons over the course of a unit of study. We hope that you will use the models we have provided in this chapter as a guide for your own planning and teaching. We encourage you to adapt them for your own students' needs.

Grades K–2 Science Lesson

The first lesson plan that follows is part of a unit for elementary students in the primary grades and is related to Earth Science, or, specifically in this case, objects in the universe. Presented here is Day 1, which introduces the students to objects in the sky, such as the sun and moon. The teacher incorporated three of the techniques from Chapters 2 and 3 for the young learners. As they consider the differences among objects that give light or do not, they classify information on a **T-Chart** (see Figure 4.7). Then at the end of the lesson, they review their ideas through a **Milling to Music** activity, which is explained to them using a **Procedures with Graphics** poster.

As you read this lesson, notice that even for these primary grade students, the lesson includes both content and language objectives. Students will be classifying objects that give light or do not, and they will be using clauses of time to tell when they see the sun and the moon. The teacher provides sentence starters to support their oral language practice and written responses.

SIOP® LESSON PLAN: *Grades K–2: Objects in the Universe, Day 1* (Developed by Amy Ditton)

Key: SW = Students will **TW** = Teacher will **HOTS** = Higher Order Thinking Skills

Lesson: Sun and Moon

Standard: SW recognize the changes that occur in a 24-hour day. SW observe and describe the changes in the position of the sun and the moon.

Key Vocabulary:	**Visuals/ Resources/Supplementary Materials:**
universe, position, light, sun, moon, sky **HOTS:** Classifying, predicting.	Key vocabulary posters with graphics T-Chart posted on chart paper T-Charts for students (one per group) (see Figure 4.7) Various picture cards of objects seen in the sky, including "sun" and "moon" (1 set per group) Procedures with graphics for Milling to Music

Connections to Prior Knowledge/Provide Background Information:

- Teacher and students will go outside to the playground or other area on the school grounds to observe the sky.
- SW work with a partner to create a list of objects they see in the sky.
- Back in the classroom, TW randomly call on students to share an observation with the whole group.
- TW remind students that they have studied different cycles. TW refer to the term "cycle" that students have used in previous lessons. (TW use a non-linguistic representation of "cycle" that students are familiar with, such as an arm movement, photo, or an illustration.)
- TW explain to students that they are going to be learning about a cycle that takes place each day in the sky.
- TW share the content and language objectives with the class.
- Using illustrations and student-friendly definitions, TW introduce the key vocabulary to the class. (Key vocabulary words with graphic illustrations will remain posted in the classroom for students to refer back to.)

Objectives:	*Meaningful Activities:*	*Review/Assessment:*
Content Objectives: SW classify different objects that they see in the sky.	• TW provide students with labeled picture cards of objects seen in the sky. (The terms "moon" and "sun" should be included.)	
Language Objective: SW use clauses of time to discuss when they have seen the sun or the moon in the sky. I saw the sun when . . . I saw the moon when. . . .	• TW model using a **T-Chart** to classify the cards into two categories—objects that give light and objects that do not give light. • SW work with a partner to classify the objects on a T-Chart.	• TW circulate the room as students are working with their partners on the T-Chart.
	• TW ask students to share items from their chart that do give light. SW share using the sentence frame: "One object that gives light is ________________."	• TW randomly call on students to share their responses.
	• TW model the language objective phrases and briefly explain clauses of time. Students will use the sentence frames to write down different times that they have seen the sun and the moon in the sky.	• TW collect students' written responses.
	• TW review the procedures for **Milling to Music** and will refer students to the "Milling to Music" **procedures with graphics** poster. SW share their responses in a "Milling to Music" activity.	• TW circulate the room as students share their responses.

Wrap Up: SW respond to the following question in their science journals: "How do you think the sun and the moon are part of a cycle?"

TW refer students back to the content and language objectives and will ask students if they met the objectives, and if so, how.

(Template developed by Melissa Castillo and Nicole Teyechea. Used with permission.)

FIGURE 4.7 *T-Chart for Objects in the Sky*

Objects That Give Light	Objects That Do Not Give Light
sun 	clouds 

Grades 3–5 Science Lesson

The second lesson plan example is part of a unit on Newton's first law of motion, a Physical Science unit for upper elementary grade students. On Day 1 of the unit, students conducted some mini-experiments. Now, at the start of Day 2 in the lesson shown here, they recall some of their observations using the **Oh Yesterday!** technique. The content objective is designed for students to apply their observations to Newton's first law of motion, and the language objective helps them phrase their observations and analyses scientifically. The teacher uses the **Web of Information** activity to check on student comprehension at the end of the lesson.

SIOP® LESSON PLAN: *Grades 3–5, Day 2: Newton's First Law of Motion* (Developed by Amy Ditton)

Key: SW = Students will **TW** = Teacher will **HOTS** = Higher Order Thinking Skills

Day 2 Lesson: Explaining the Mini-Experiments

Standard: Identify the conditions under which an object will continue in its state of motion (Newton's 1st Law of Motion).

Key Vocabulary:
object, rest, motion, force, gravity, friction

HOTS: Students analyze Mini-Experiments and synthesize their observations with Newton's First Law of Motion.

Visuals/Resources/Supplementary Materials:
Vocabulary words with graphics posted in classroom

Chart paper with Newton's First Law of Motion written on it and posted in the room

Science journals

Mini-Experiments Summary Chart (see Figure 4.8)

Ball of yarn or string

Connections to Prior Knowledge/Provide Background Information:
- TW introduce content and language objectives.
- TW ask students to perform **Oh Yesterday!** to relate at least one observation for each of the mini-experiments from the previous day. Students may refer back to their notes and science journals as they recall their observations. TW show the digital photos from the day before (if they were taken).

Objectives:	*Meaningful Activities:*	*Review/Assessment:*
Content Objective: SW be able to apply Newton's Law of Motion to their observations from the mini-experiments.	• TW use the Mini-Experiments Summary Chart (Figure 4.8) to model applying Newton's First Law of Motion to one observation about a mini-experiment.	
	• SW work with the same small groups from the previous lesson to apply Newton's First Law of motion to their observations of the three mini-experiments. Each student will record his/her responses in the boxes of the top half of the summary chart.	• TW circulate the room asking groups questions and providing feedback as students work.
Language Objective: SW use key vocabulary and sentence frames to explain their observations from the mini-experiments. __________ stayed at rest because . . . __________ was in motion. __________ was the force that acted on . . .	• SW share their responses in a "Numbered Heads Together" activity. • TW next model taking a description from the summary chart boxes and explaining the description using key vocabulary and sentence starters. SW complete the bottom half of the summary chart sheet.	• TW assess student responses in the Numbered Heads Together activity. • TW collect summary charts to read student responses.

Wrap Up:
- SW share their explanations using the **Web of Information** activity. The class will sit in a circle and as each student shares an explanation, he or she hangs on to a piece of yarn or string before tossing the ball to the next speaker.
- TW refer students back to the content and language objectives and will ask students if they were met, and if so, how.

(continued)

FIGURE 4.8 *Mini-Experiments Summary Chart*

An object at rest stays at rest

AND

an object in motion stays in motion

UNLESS

a force acts upon it.

Mini-Experiment	Running	Flick the Note Card	Pile of Pennies
Observation	I fell when I stopped.		
Object at Rest			
Object in Motion	Me		
Force	Friction with my feet when I tried to stop.		

Sentence Frames:

__________ **stayed at rest until** . . .
__________ **was in motion until** . . .
__________ **was the force that acted on** . . .

Running Explanations:
I was in motion until I used my feet to try to stop.
Friction was the force that acted on me.

Flick the Note-Card Explanations:

Pile of Pennies Explanations:

Grades 6–8 Science Lesson

The third lesson is for middle school students in a Physical Science class, typically taught in Grade 7 or Grade 8. It falls as Day 4 in a unit on the rock cycle. You will notice that a **Quickwrite** activity is used for building background at the start of the lesson. Next **4-Corners Vocabulary Charts** are modeled by the teacher and then developed with new vocabulary words by the students. The students review the information they have learned through the **Milling to Music** technique.

Besides these activities, you will see the content and language objectives that the teacher designed as well as the higher order thinking skill she expects the students to practice, namely applying vocabulary to new contexts. All of these activities and thinking processes take place in connection to the rock cycle.

SIOP® LESSON PLAN: *Grades 6–8, Day 4: Science Rocks!*

(Developed by Hope Austin-Phillips)

Key: SW = Students will **TW** = Teachers will **HOTS** = Higher Order Thinking Skills

Day 4 Lesson Title: The Rock Cycle Part I

Content Standards:
Describe the rock cycle.
Distinguish the components and characteristics of the rock cycle for the following types of rocks:
- igneous
- metamorphic
- sedimentary

Key Vocabulary:
rock cycle, metamorphism, melting, crystallization, weathering, transportation, deposition, compaction, cementation, pressure

HOTS: Applying vocabulary in context

Visuals/Resources/Supplementary Materials:
Igneous, sedimentary, and metamorphic rocks

Projected image of the rock cycle diagram (from textbook or Internet)

Key vocabulary words with graphics or illustrations on cards

4-Corners Vocabulary Charts

Animation of rock cycle processes (from textbook or Internet)

Connections to Prior Knowledge/Building Background:
SW do a 2-minute **Quickwrite** to relate what they have already learned about rocks. SW pair-share their Quickwrite. TW use magic name sticks to call on students to report out. TW review the objectives and tell students that today they will be learning about the rock cycle. TW explain that there are some processes that make the rocks change into other kinds of rocks and they are going to be learning about that. TW hold up a sedimentary rock that the students have seen before and tell the students that at some point the sedimentary rock might become a metamorphic rock or igneous rock. TW hold up one of these other types of rocks for students to see as a reminder of what the different rocks look like.

Objectives:	*Meaningful Activities:*	*Review/Assessment:*
Content Objective: SW explore the stages of the rock cycle. **Language Objectives:** SW draw an illustration and write a contextualized sentence to show understanding of the rock cycle vocabulary words. SW orally share definitions of key words with partners.	• TW hand out a **4-Corners Vocabulary Chart** to each student. TW introduce each vocabulary word (e.g., *weathering*) by saying the word, explaining what it means, showing a graphic representation of the word, and showing an animation if there is one available. • TW write one word and a definition for the word on the 4-Corners Vocabulary Chart. SW copy the word and definition onto their 4-Corners Vocabulary Chart (or they can draw the chart in their notebook). TW model how to draw a graphic representation of the word and will model how to write a contextualized sentence for the first word. SW copy the teacher's example.	

(continued)

SIOP® LESSON PLAN: *Grades 6-8, Day 4: Science Rocks!*

(Developed by Hope Austin-Phillips) *(continued)*

Objectives:	*Meaningful Activities:*	*Review/Assessment:*
	• TW write a second word and a definition. SW copy the word and definition. SW draw an image of the word and will write a sentence for the word. TW call on volunteers to share their drawings and sentences. TW give specific feedback to correct any errors. • The process will continue until all of the words are completed. (See Think-Aloud 1) • TW tell students that in the rock cycle, rocks can change into other kinds of rocks. TW project an image of the rock cycle for the students to see. (See Planning Point 1.) TW show students where sedimentary, igneous, and metamorphic rocks are on the diagram. • TW think-aloud to draw and label a rock cycle diagram that students copy in their notebooks. Together they label their diagrams with key vocabulary.	• TW assess students by monitoring their work on the 4-Corners Vocabulary Charts.
	• TW remind students of the second language objective and review directions for **Milling to Music**. SW take their 4-Corners Vocabulary Charts with them during the activity. Each time the music stops, TW ask students to share a vocabulary word and definition with their partner.	• During Milling to Music, TW monitor student responses.

Wrap-Up: Students return to their seats. TW review the objectives for the day and will ask students if they have met them. TW name some key words and SW think-pair-share at least one definition. TW use magic name sticks to call on students to share a response.

TEACHER THINK-ALOUD 1

If time is short, I may organize students into groups and assign different words to different students in the groups. They complete the 4-Corners Vocabulary Charts for their words and then share with group members.

PLANNING POINT 1

It is useful to show animations of the rock cycle on the computer. The following site has two Web pages: http://www.classzone.com/books/earth_science/terc/content/investigations/es0602/es0602page02.cfm and http://www.classzone.com/books/earth_science/terc/content/investigations/es0602/es0602page03.cfm

Grades 9–12 Science Lesson

The final lesson in this chapter is drawn from a high school biology unit on the cell. In this fourth day lesson, you will see how the teacher taps the students' prior knowledge about structures in plant and animal cells using the **Snowball!** activity as a warm-up. The teacher also incorporates the **Signal Words** poster to help students express comparisons and contrasts between animal and plant cells. These two activities support the content and language objectives that ask students to use comparative and contrastive language structures.

SIOP® LESSON PLAN: *Grades 9–12, Day 4: Cells*

(Developed by Hope Austin-Phillips)

Key: SW = Students will **TW** = Teachers will **HOTS** = Higher Order Thinking Skills

Day 4 Lesson: Comparing and Contrasting Plant and Animal Cells
Content Standard(s): Compare and contrast plant and animal cell structures

Key Vocabulary:	**Visuals/Resources/Supplementary Materials:**
animal cell, plant cell, cell wall, cell membrane, cytoplasm, nucleus, nuclear membrane, nucleolus, mitochondria, golgi apparatus, rough endoplasmic reticulum, smooth endoplasmic reticulum, ribosomes, lysosomes, chloroplasts **HOTS:** Comparing; contrasting	Vocabulary words with graphics Snowball form (Figure 4.9) Venn diagram (See Planning Point 1.) Comparative Signal Words poster (Figure 4.10) Chart paper with two columns, headed "Structure" and "Function" posted on board or wall

Connections to Prior Knowledge/Building Background:
Snowball! SW write facts they have learned about plant and animal cells on their Snowball form (Figure 4.9). (See directions for Snowball! in Chapter 3.) After the Snowball activity, TW use magic name sticks to randomly call on students to share what they have learned. TW review the objectives for the day.

Objectives:	*Meaningful Activities:*	*Review/Assessment:*
Content Objectives: SW visually represent plant and animal cell structures. SW categorize information about the structures and functions of the two types of cells using a Venn diagram.	• TW refer back to the lab activity in which the students observed both human cheek cells and onion skin cells. TW explain that today, they are going to be looking deeper into the similarities and differences of the animal and plant cells. • SW draw diagrams of an animal cell and a plant cell and label the cell organelles. Then they complete a Venn Diagram to compare and contrast plant and animal cells.	• TW check on the students' labeled drawings of animal and plant cells and their Venn Diagrams.
Language Objective: SW use comparative language terms and phrases to state the similarities and differences between animal and plant cell structure and function to a partner.	• TW review and model comparative language with the students using the Compare/Contrast Signal Words poster. (Figure 4.10) SW think-pair-share the similarities and differences between the cheek cells and onion skin cells from the lab and try to use the signal words and sentence frames, such as	• TW use magic name sticks and have students share some sentences.

(continued)

SIOP® LESSON PLAN: *Grades 9–12, Day 4: Cells*

(Developed by Hope Austin-Phillips) *(continued)*

Objectives:	*Meaningful Activities:*	*Review/Assessment:*
	One similarity between animal and plant cell structure/function is ______. One difference between animal and plant cell structure/function is _____. • After oral practice, SW write three comparative sentences on a sticky note. SW read their sentences to a partner.	• SW place their sticky notes on the poster under the category of "structure" and "function" as appropriate.

Wrap-Up: TW review the goals for the day. SW raise one finger if they do not think that they mastered the goals for the day. SW raise two fingers if they partially understand the goals for the day but need a little more practice. SW raise three fingers if they feel they fully mastered the content and the comparative language structures.

(Template developed by Melissa Castillo & Nicole Teyechea. Used with permission.)

PLANNING POINT 1

The teacher could provide an outline of a Venn diagram to every student. However, this graphic organizer is relatively easy for students to draw on their own, as is done in this lesson.

FIGURE 4.9 *Snowball Form*

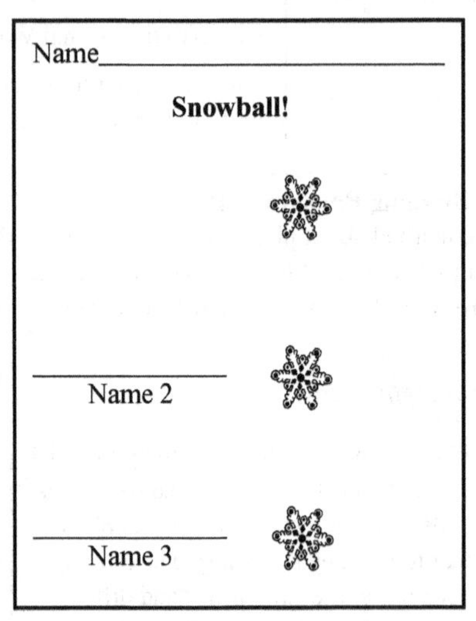

Name______________________

Snowball!

____
Name 2

____
Name 3

FIGURE 4.10 *Signal Words for Compare/Contrast Poster*

Signal Words

If you are asked to Compare
use these words:

both
alike
the same as
similar to
also
neither
in comparison

If you are asked to Contrast
use these words:

one but not the other
different from
differs
unlike
in contrast
however

Concluding Thoughts

This chapter explores lesson and unit planning for SIOP® science classes and offers insights into the thinking processes involved. We have outlined how to effectively incorporate techniques and activities into thoughtful, standards-based SIOP® science lessons that give attention to language development every day. The four lesson plans demonstrated how teachers can create opportunities for students to process content concepts and practice the academic language that supports the content. For English learners to make adequate progress in science—and all subjects—they must be given opportunities to use English for academic purposes. It is impossible to separate out the process of understanding and learning scientific concepts, theories, hypotheses, processes, and procedures from their language requirements.

A prime goal for this chapter has been to demonstrate how the techniques can be added to science lessons. In the following chapters, you will hear from SIOP® science educators as they describe more details about the specific planning processes they undertook while developing and implementing the units. We hope that these units and lessons will illustrate clearly to you how to implement the features of the SIOP® Model in daily lessons, consistently and cohesively throughout a unit of study.

Revisiting Mrs. Allen

Mrs. Allen has worked with a team of science teachers on a weekly basis to incorporate the SIOP® Model into her lesson plans. She hasn't given up discovery-based learning, but has helped students focus their discoveries on the content she is teaching. At one weekly meeting, Mrs. Allen reflected on the process: "I've learned to write objectives in a way that doesn't tell students what they're going to discover exactly, but rather, discusses their learning of a scientific concept. I've realized that with English learners, it is necessary to spend time before the discovery activity developing background knowledge. When I take the time to build background, it is a way to scaffold the students to higher levels of scientific discovery. For example, in a lesson about the different types of rocks, I give the students specific information about how to observe the rocks and how to perform tests on the rocks, but I do not tell them what they will discover. This way, they will see that acid reacts with limestone rocks and they will see that igneous rocks have crystals. If they discover this before they learn about the different types of rocks, they will then be able to incorporate this experiential background knowledge into their understanding of the rock cycle."

When a colleague asked her how she was handling the language development side of the SIOP® Model, Mrs. Allen replied, "I've also focused on developing their vocabulary and oral discussion skills. Sometimes this means teaching the vocabulary words before the learning takes place, but sometimes I wait until after they've made some discoveries so they have a concrete experience to connect the abstract terms to. I'm also teaching them some sentence frames to help them discuss their observations and conclusions."

"I think I would sum it up in this way. With my procedures, I'm giving the students enough information so that they understand what they need to do with the materials. I'm not telling them what their discovery will be, but I'm providing scaffolds that advance the students to a place where they can make scientific discoveries. Without the scaffolding and background building, they don't know what they might be discovering nor how to discuss it, so the value of the discovery is nonexistent. It's rewarding to see that students are making observations on their own, but they can now connect the observations to grade-level scientific concepts."

chapter 5

SIOP® Science Units and Lessons for Grades K–2

Amy Ditton and Deborah Short

Introduction

This chapter and the three that follow present complete five-day science units, including lesson plans and student handouts. In this chapter, SIOP® science educator Amy Ditton discusses in detail her planning process for a unit for students in the primary grades. She also builds on the lesson planning discussion from Chapter 4 by walking us through her decisions and selections when writing the Day 2 lesson. (*Note*: If you have not read Chapter 4 about SIOP® science unit and lesson design for English learners, please read it for an overview of the planning process and a discussion of the lesson plan format used in this and successive chapters.) As you read this unit, look at how several of the techniques and activities from Chapters 2 and 3 are incorporated into the lessons! See how the activities in the unit lesson plans map onto the components of the SIOP® Model (found in Appendix A).

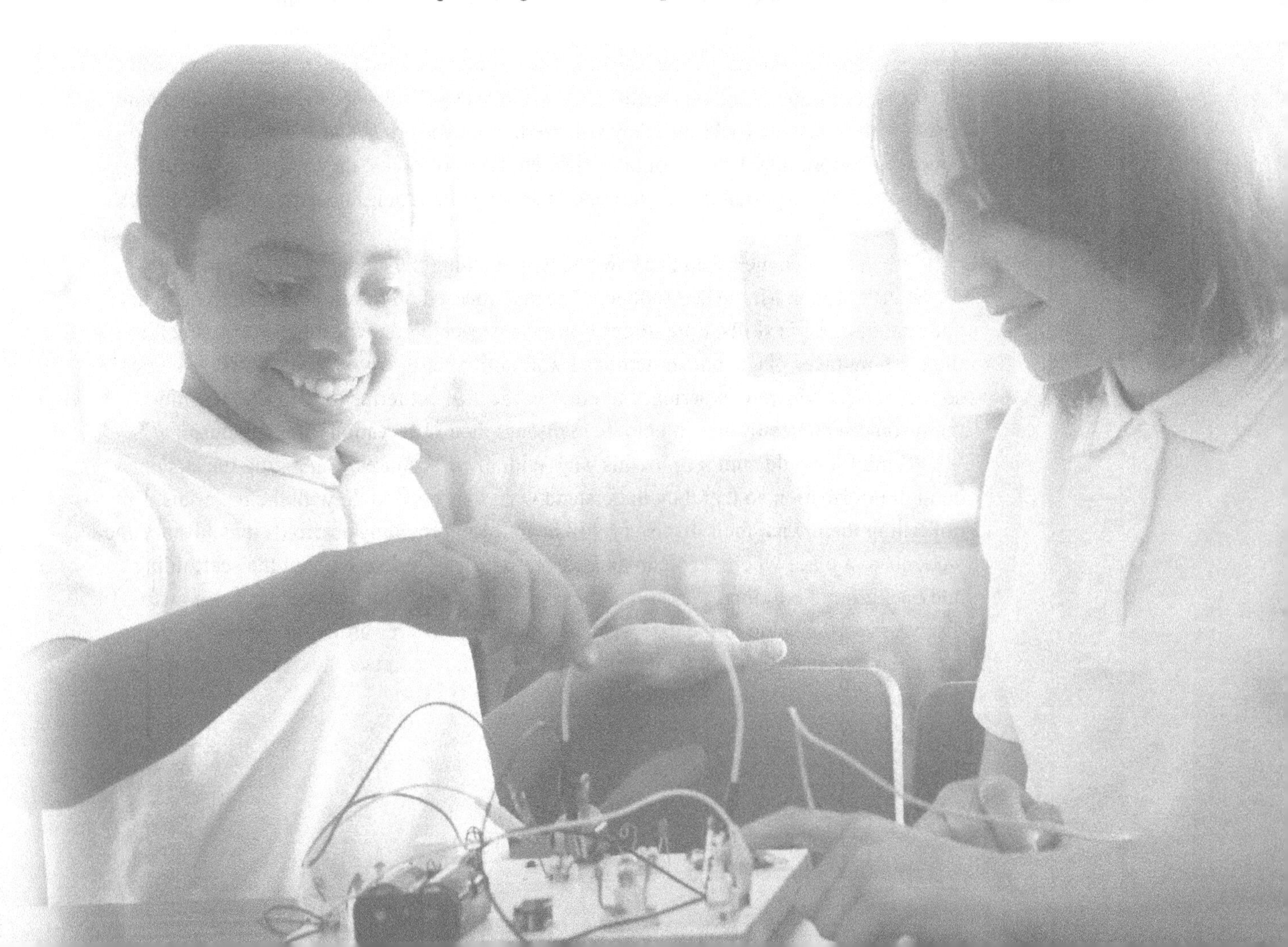

You will notice that there are teacher Think-Alouds and Planning Points in many of the lessons. The Think-Alouds, identified with a thought bubble icon, are questions we asked ourselves and/or decisions we made as we undertook the SIOP® unit-planning process. The purpose of the Planning Points, identified with a flash drive icon, is to highlight or identify resources that support the implementation of the lesson.

Objects in the Universe

Unit Overview, Grades K–2

It is a lot of fun teaching science to young learners. They are so curious about the world! When I initially outlined this unit, I started with just a sequence of content and language objectives and some grouping configurations that I wanted to include. In my outline, I didn't have a list of what students would do during different times of the day, and there wasn't a student-created book. But I knew that by the last lesson, I wanted the students to explain what time of day was their favorite.

As I began drafting lesson plans, I realized that I was not including enough questions or tasks that required students to use higher order thinking skills in the initial lessons, so I revised my lessons. Some of the activities for Lessons 3 and 4 became incorporated into Lessons 1 and 2. I reworked Lessons 3 and 4 to advance students to higher levels of understanding. After double-checking the science standards, I also decided that this unit was best for Grades 1 or 2.

The first lesson (Day 1) heavily emphasizes Building Background for learning about the sun and moon. In this lesson, students are making connections to their prior experiences, and there is an introduction to key vocabulary for the unit. Students are also making connections to past learning by referring back to the concept of a "cycle." I chose to link to previous learning with the concept of "cycle" because there are state standards that refer to plant cycles, animal cycles, and weather cycles. Thus this academic concept is highly utilized.

I wanted to focus on classification, so I thought at first about having students classify the objects that they saw. However, I realized that students would have more objects to discuss and sort if I provided them. In other words, if the students are outside on the morning of the lesson, the only object that they will see in the sky that gives light will be the sun. So I augmented the lesson with labeled picture cards. By providing cards for students to sort, other examples could be offered, such as the moon and stars. I wanted to be sure that all students were introduced to the sun and moon as two objects in the sky that provide light. The labeled picture cards were valuable for another reason—they would assist beginners in naming familiar objects.

On Days 2 and 3, students learn about identifying the different phases of a 24-hour day. My goal for the students was to ensure that they understand what happens at different times of the day and how these times of day relate to their lives. To promote language practice and tap higher order thinking, I planned for students to explain which phase their "favorite time of day" falls in as a warm-up for Day 3. I wanted to point out that it is important for teachers to assess whether student responses seem appropriate for each time of day. If a response seems unusual, the teacher should find out if there is a misconception, or if the child is sharing an answer that makes sense to himself or herself. For example, one student might stand up to demonstrate that his favorite time of day is when the sun comes up, or "sunrise." The teacher may ask the student to explain his reason for choosing that as his favorite time of day and the response could be, "That's when my dad gets home from work." The teacher might need to find out more. Does the child's father work a night shift, or is the child confusing "sunrise" with "sunset"?

I would have liked to do more with a sun/moon journal that would have students observing the sun/moon outside. However, depending on the time of year the lesson is being taught and the region where students live, it may or not be possible to have young students outside at night looking at the moon.

On Day 3, students shift from relating one event to a certain time of day to relating many events to different times of the day. For young learners, moving from a smaller concept to a larger one is beneficial. I added to the lesson plan the Think-Pair-Share activity that requires students to explain different reasons why they related a picture to a certain time of day. This was another opportunity to include higher order thinking, as well as a way to help the teacher determine whether some students have misconceptions or different perspectives. If a student writes "sunset" for the picture of the rooster on the chart, for instance, there could be a reason that does not indicate a misconception. The teacher may be looking for the response "sunrise," but if asked why she chose "sunset," the child may explain that every evening she puts their rooster in its shed. With our culturally diverse students, we should always be open to different perspectives and not make assumptions.

I designed the Day 4 lesson to evaluate student learning and comprehension of the state standards that were addressed throughout the first three days. In planning this lesson, I wondered if it would be a good use of instructional time to have students drawing because the task is not part of the state standard. I decided nonetheless that it would be appropriate because the flip-book promotes student synthesis of the concepts and provides the teacher with an authentic piece of student work to evaluate. It also gives students an additional opportunity to meaningfully practice and apply both the content and the language of the lesson. As the teacher looks at the student work that results from this lesson, he or she can look for more detailed drawings for each phase during the day. The teacher can also assess how a student is using scientific language. As I realized that students may need more than one day to complete the flip-books, I noted that this final lesson may take two days to complete.

Day 2 Lesson Design: Changes Throughout the Day

This unit focuses on the standard that states: "Student will recognize the changes that occur in a 24-hour day. The student will observe and describe the changes in the position of the sun and the moon." As mentioned earlier, the first day of the lesson focuses on building background: helping students identify the objects they will be talking about and having students identify different times that they have seen the sun and the moon in the sky.

It is in Day 2 of this unit that students really begin to directly work toward the standard. Since the SIOP® component Building Background is heavily emphasized in Day 1, students are ready to move right to the content topic related to the standard with just a quick review of background experiences and past learning that will support them in learning the new content. New vocabulary is introduced as the substance of the lesson.

The content objective for the lesson is: Students will identify changes that occur in the sky during a 24-hour period. The language objective is: Students will use sentence frames with time clauses to describe the changes in the position of the sun and the moon, such as "When the ___(sun/moon)_____ is ___(rising, at the top of the sky, setting)___, it is __(sunrise, midday, sunset, moonrise, midnight, moonset)___." Together, the content and language objectives scaffold the process so students can directly meet the state standard.

The first thing I wanted students to do was see the connection between the cycle of changes throughout the day and other cycles they had studied. There were many different examples of types of cycles that first and second graders study, so I thought an easy connection to past learning would be to have students think about the other cycles they are familiar with. (This was something they thought and wrote about on Day 1, so they could quickly refer back to it.) This connection helps them link the prior academic concepts with "the changes in a 24-hour period."

The next step of the lesson is for the teacher to introduce and give student-friendly definitions for each of the phases. The teacher will need to supplement the definitions with photos so that students who are not proficient in academic English will have visual support. I strongly suggest that the teacher use actual photographs taken in an environment similar to where the students live. For example, on the West Coast, sunrise doesn't happen over the ocean, sunset does, but a photo from the East Coast would show the opposite. Sunrise and sunset could look very similar, so the teacher will need to select photos carefully. Having students go outside throughout the day will also concretize their understanding of the concepts of *sunrise, midday*, and *sunset*. Depending on the time of year, the teacher may be able to ask some students to go outside in the evening to look at the sky as well.

In the first draft of this lesson plan, I also included opportunities for students to discuss their favorite time of day as the teacher introduces the different phases. After reviewing the plan, I decided to remove the activity because it detracted from the learning goal. Instead, I used it as a warm-up activity for the next day.

Through the language objective, students have the opportunity to discuss and write about each phase of the 24-hour cycle. The language use is very scaffolded through the use of sentence frames. If students are more proficient with academic English, however, they may not need the frames. The teacher can determine whether the use of frames is appropriate for all students, for some students, or is unnecessary.

After students write about each phase of the 24-hour cycle, they get to share their responses in an Inside-Outside Circle activity. There are many different ways that students could share, but I chose Inside-Outside Circle because it is one of my favorites. One reason I like using Inside-Outside Circle is that it gives students multiple opportunities to share their responses in a short amount of time. I also like that it offers a structure in which the teacher can have more proficient students share first, so that less proficient students have a chance to listen before they speak. Students don't need to know that it is set up that way. For example, a teacher can have the more proficient students assigned to be in the Inside Circle while the less proficient students are in the Outside Circle. The teacher can then prompt the Inside Circle to share first, allowing less proficient students the opportunity to listen before sharing. The other reason I like facilitating Inside-Outside Circle is that I can stand in the middle of the circle and monitor everything that students are saying.

Like many other Interaction techniques, Inside-Outside Circle can take some time to set up. The first time I used it in my classroom, it took students 10–15 minutes to set up the circles and understand the procedures for sharing and rotating. The second time we tried it, it moved more quickly, and the third time we did Inside-Outside Circle, students knew how to get into the circles expeditiously and began to share promptly. The time spent up front teaching the routine paid off in opportunities for students to share.

The final activity in the lesson is a Think-Pair-Share in which students quickly refer back to their responses from the previous lesson. On Day 1, students were asked to name a time they had seen the sun or the moon. To connect to the new information, students now have to identify what phase the sun or moon was in when they saw it.

K–2 Unit

Now you will have a chance to read the Objects in the Universe unit and to see how Amy Ditton organized her planning into interactive SIOP® science lessons for the students. Some student worksheets are listed as BLMs (Blackline Masters) and are located in Appendix C.

SIOP® LESSON PLAN: *Grades 1–2 Unit: Objects in the Universe, Day 1: Sun and Moon* Developed by Amy Ditton

Key: SW = Students will **TW** = Teacher will **HOTS** = Higher Order Thinking Skills

Standard: SW recognize the changes that occur in a 23-hour day. SW observe and describe the changes in the position of the sun and the moon.

Key Vocabulary:
universe, position, light, sun, moon, sky

HOTS: Classifying, evaluating

Visuals/Resources/Supplementary Materials:
Key vocabulary posters with graphics

T-Chart posted on chart paper

T-Charts for students (one per pair) (BLM #1)

Various picture cards of objects seen in the sky, including sun, moon, clouds, birds, stars, airplane, and so on (1 set per pair)

Procedures with graphics for Milling to Music (Figure 5.1)

Connections to Prior Knowledge/Provide Background Information:

- Teacher and students will go outside to the playground or other area on the school grounds to observe the sky.
- SW work with a partner to create a list of objects they see in the sky.
- Back in the classroom, TW randomly call on students to share an observation with the whole group.
- TW remind students that they have studied different cycles. TW refer to the term "cycle" that students have used in previous lessons. (TW use a non-linguistic representation of "cycle" that students are familiar with, such as an arm movement, photo, or an illustration.)
- TW explain to students that they are going to be learning about a cycle that takes place each day in the sky.
- TW share the content and language objectives with the class.
- Using illustrations and student-friendly definitions, TW introduce the key vocabulary to the class. (Key vocabulary words with graphic illustrations will remain posted in the classroom for students to refer back to.)

Objectives:	*Meaningful Activities:*	*Review/Assessment:*
Content Objective: SW classify different objects that they see in the sky.	• TW provide students with labeled picture cards of objects seen in the sky. (The terms "moon" and "sun" should be included.) • TW model using a T-Chart (BLM #1) on chart paper to classify the cards into two categories—objects that give light and objects that do not give light.	
Language Objective: SW use clauses of time to discuss when they have seen the sun or the moon in the sky: I saw the sun when . . . I saw the moon when. . . .	• SW work with a partner to classify the objects on a T-Chart. • TW ask students to share items from their chart that do give light. Students will share using the sentence frame: "One object that gives/does not give light is ____________."	• TW circulate the room and check as students are working with their partners on the T-Chart. • TW randomly call on students to share their responses.
	• TW model the language objective phrases and briefly explain clauses of time (e.g., "I saw the sun when I was driving to school." "I saw the moon when I walked outside of the movie theater at night."). SW use the sentence frames to write down different times that they have seen the sun and the moon in the sky.	• TW collect and review students' written responses.
	• TW review the procedures for "Milling to Music" (see Chapter 3) and refer students to the "Milling to Music" procedures poster (Figure 5.1). • SW share their responses in a "Milling to Music" activity.	• TW circulate the room as students share their responses.

Wrap Up: SW evaluate the sun and moon cycle by responding to the following question in their science journals: "How do you think the sun and the moon are part of a cycle?"

TW refer students back to the content and language objectives and will ask students if they met the objectives, and if so, how.

(Template developed by Melissa Castillo and Nicole Teyechea. Used with permission.)

FIGURE 5.1

Milling to Music Procedures

When the music begins, walk slowly around the room.

When the music stops, find a partner standing close to you and raise your hands.

At my signal, share your response with your partner.

When the music begins again, the process starts over.

SIOP® LESSON PLAN: *Grades 1–2 Unit: Objects in the Universe, Day 2: Changes Throughout the Day, Part I*

Developed by Amy Ditton

Key: SW = Students will **TW** = Teacher will **HOTS** = Higher Order Thinking Skills

Standard: SW recognize the changes that occur in a 24-hour day. SW observe and describe the changes in the position of the sun and the moon.

Key Vocabulary:
sunrise, midday, sunset, moonrise, midnight, moonset, rising, middle, setting

HOTS: Summarizing, determining importance

Visuals/ Resources/Supplementary Materials:
Sun/Moon Phases Chart (side 1 of BLM #2) – one copy per student. *Note*: make double-sided copies of the Sun/Moon Phases Chart

Large Sun/Moon Phases Chart (on chart paper or transparency)

Photos/illustrations of the sun and moon in the sky throughout the day

Science journals

Connections to Prior Knowledge/Provide Background Information:
- TW refer students back to their science journals and ask them to share their responses from the previous lesson. ("How do you think the sun and the moon are part of a cycle?")
- TW explain that the sun and moon change position throughout the day, every day, creating a cycle.
- TW introduce the content and language objectives for the lesson.
- TW introduce new key vocabulary words (*sunrise, midday, sunset, moonrise, midnight, moonset, rising, middle, setting*) and review vocabulary words (*universe, position, light, sun, moon*) from the previous lesson. New vocabulary words will also be posted with illustrations for students to refer back to.

Objectives:	*Meaningful Activities:*	*Review/Assessment:*
Content Objective: SW identify changes that occur in the sky during a 24-hour period. **Language Objective:** SW use sentence frames with time clauses to describe the changes in the position of the sun and the moon: When the (sun/moon) is (rising, at the top of the sky, setting), it is (sunrise, midday, sunset, moonrise, midnight, moonset).	• Each student will receive a Sun/Moon Phases Chart divided into 6 sections—each representing a time of day. (See BLM #2.) Teacher will have a large graphic organizer at the front of the room for modeling. (See Think-Aloud 1.) • Using photos, the teacher will introduce the concept of "sunrise" to the students. • As the teacher introduces each phase to students, he/she will point out the position of the sun in the sky. The photos the teacher uses should show whether the sun or moon is rising, at the top of the sky, or setting. (Phases include: sunrise, midday, sunset, moonrise, midnight, moonset.) • TW model drawing a picture and writing a definition of "sunrise" in the first square of the graphic organizers. • SW each draw a picture of sunrise and write a definition in the first square of their graphic organizers. • Teacher and students will repeat the procedures for the concepts of midday, sunset, moonrise, midnight, and moonset.	• TW circulate and check as students work to fill out charts. • TW collect and review charts.

• SW individually use the sentence starters to write six sentences describing changes that happen in the sky throughout the day. • SW share their responses with an "Inside-Outside Circle" technique.	• TW listen to student responses and will ask questions to encourage students to elaborate on their response. • TW stand in the middle of the Inside-Outside Circle to listen to student responses. • TW collect student responses as "tickets out."

Wrap Up:

- TW refer back to the content and language objectives and will ask students if they were met, and if so, how.
- SW refer back to the statements they wrote on Day 1. ("I saw the sun when . . . , I saw the moon when . . .") SW identify what time of day it was when they saw the sun or moon, and will share in a "Think-Pair-Share" activity. After thinking about it, a student might say, "I saw the moon when I came out of the movie theater. The phase was moonrise."
- Throughout the school day, SW go outside to look at the sun and note in their science journals where the sun is positioned in the sky (e.g., rising, at the top, setting).

(Template developed by Melissa Castillo and Nicole Teyechea. Used with permission.)

THINK-ALOUD 1

I think it is always important to model how to use the charts and other graphic organizers for young learners. It's never too early to start learning how to record information and take notes!

SIOP® LESSON PLAN: *Grades 1–2 Unit: Objects in the Universe, Day 3: Changes Throughout the Day, Part II*

Developed by Amy Ditton

Key: SW = Students will **TW** = Teacher will **HOTS** = Higher Order Thinking Skills

Standard: SW recognize the changes that occur in a 24-hour day. SW observe and describe the changes in the position of the sun and the moon.

Key Vocabulary:
universe, position, light, sun, moon, sky, sunrise, midday, sunset, moonrise, midnight, moonset

HOTS: Justifying

Visuals/Resources/Supplementary Materials:
Sun/Moon Phases Chart (side 2 of BLM #2)
Photos/illustrations of the sun and moon in the sky throughout the day
Procedures with graphics
Science journals
Teacher-created slide show
Whiteboards

(continued)

SIOP® LESSON PLAN: *Grades 1–2 Unit: Objects in the Universe, Day 3: Changes Throughout the Day, Part II*

Developed by Amy Ditton *(continued)*

Connections to Prior Knowledge/Provide Background Information: TW redistribute to students the Sun/Moon Phases Charts they completed in the previous lesson. SW share favorite time of day and explain which phase it falls in during a "Milling to Music" activity (see procedures in Chapter 3). (See Think-Aloud 1.)

Objectives:	*Meaningful Activities:*	*Review/Assessment:*
Content Objective: SW relate changes that occur in the sky during a 24-hour period to other familiar events that occur during different times of the day.	• TW show photos of different events that take place throughout the day (e.g., rooster crowing, children sleeping, family eating dinner). • As the teacher shows each picture, he/she will model a think-aloud, asking, "What time of day does this happen?" • SW think about the time of day that the event occurs and write their answer on a whiteboard.	
Language Objective: SW list and discuss events that occur during different times of the day.	• SW hold their answer up for the teacher to see (e.g., the photo of a rooster prompts students to write "sunrise" on their whiteboard). • SW "think-pair-share" their reasons for choosing particular answers. (See Think-Aloud 2.) • TW use side 2 of the Sun/Moon Phases Chart (BLM #2) to model how to list things that he or she does at different times during that day (e.g., sunrise—wake-up, drink coffee). • SW use the back of their charts to list different events that occur during different times of their day (e.g., midnight—sleep, dream). • SW share their responses in a "Pass the Notecard" activity (see procedures in Chapter 3).	• As students write responses on their whiteboards, teacher can informally assess comprehension. • TW listen to responses as to why students chose certain times of day in the think-pair-share activity. • TW listen to student responses in the "Pass the Note Card" activity. • TW collect charts to review student responses.

Wrap Up: TW refer back to content and language objectives and SW think-pair-share ways they can identify that the objectives were met.

(Template developed by Melissa Castillo and Nicole Teyechea. Used with permission.)

THINK-ALOUD 1

It is important for teachers to listen closely to the students during this activity to assess whether they seem appropriate for each time of day.

THINK-ALOUD 2

I like to include tasks like this because they allow for higher order thinking as well as a more accurate assessment of student comprehension. In this case, students practice the thinking skill of justification.

SIOP® LESSON PLAN: *Grades 1–2 Unit: Objects in the Sky, Day 4: Changes in the Sky* Developed by Amy Ditton

Key: SW = Students will **TW** = Teacher will **HOTS** = Higher Order Thinking Skills

Standard: SW recognize the changes that occur in a 24-hour day. SW observe and describe the changes in the position of the sun and the moon.

Key Vocabulary:
universe, position, light, sun, moon, sky, sunrise, midday, sunset, moonrise, midnight, moonset

HOTS: Summarizing

Visuals/Resources/Supplementary Materials:
Sun/Moon Phases Chart (BLM #2)

Photos/illustrations of the sun and moon in the sky throughout the day

Science journals

Flip books (See Planning Point 1 and Think Aloud 1.)

Framed summary outline (Figure 5.2)

Connections to Prior Knowledge/Provide Background Information: TW redistribute to students the Sun/Moon Phases Chart they completed in the previous lesson. SW share what they learned about each time of the day in a "Milling to Music" activity (see procedures in Chapter 3).

Objectives:	*Meaningful Activities:*	*Review/Assessment:*
Content Objective: SW describe the position of the sun and the moon at different times of day. **Language Objective:** SW use sentence starters to write about each position of the sun and moon in a flip book. The first part of the day/night is . . . The next part of the day/night is . . . The last part of the day/night is . . . When it is ________, ________ happens. When it is ________, I am _______.	• TW provide each student with a flip book (see Think-Aloud 2) with at least 8 blank pages. • TW model writing the name of each sun/moon phase at the top of 6 blank pages in the flip book. (First and last page will be left blank.) TW model leaving space for a drawing and writing on each page. • TW model using the sun and moon charts and the science journal notes from previous lessons to create a new, more detailed drawing for sunrise. SW use their own charts and notes to create a new, more detailed drawing for sunrise. • TW model using the chart and the science journal notes from previous lessons to create a new, more detailed drawing for midday. SW use their own charts and notes to create a new, more detailed drawing for midday. • SW work to use their charts and notes to draw a representation for the remaining sun and moon phases. • After the drawings are completed, TW model how to write about sunrise using at least two sentence starters. • SW use at least two sentence starters to write about sunrise. • TW repeat modeling, as necessary. • SW continue to use sentence starters to write about each phase of the day.	• TW circulate the room to informally assess and offer feedback on student work as they work independently. • As each page is completed, TW ask students to share their responses in a think-pair-share activity.

(continued)

SIOP® LESSON PLAN: *Grades 1–2 Unit: Objects in the Sky, Day 4: Changes in the Sky* Developed by Amy Ditton *(continued)*

Objectives:	*Meaningful Activities:*	*Review/Assessment:*
	• SW write a title on the first page and will write a summary of their favorite time of day on the last page (see Think-Aloud 3). If needed, students may use the "Framed Outlines" (see Figure 5.2 and procedures in Chapter 2).	• TW collect and review student-created flip books.

Wrap Up: SW read their flip books to different partners in a Conga Line (see Planning Point 2) activity. They line up in two rows facing each other. They take turns sharing their books with their facing partner. At the teacher's signal, one line takes a step to the right. The process repeats with the new partners. The last student at the end of the line that moves walks to the front of the same line to meet a new partner.

(Template developed by Melissa Castillo and Nicole Teyechea. Used with permission.)

THINK-ALOUD 1

Depending on the grade level and proficiency level of the students, this lesson may extend for two days in order to complete the flip books.

THINK-ALOUD 2

I like having students make their own flip books with pictures and key words. It allows them to remember the words better and then share the books, reading it to classmates. Also this is a nice closure activity for the unit.

THINK-ALOUD 3

For the final page of the flip book, some students may still need scaffolding in order to write their summary. One idea is to provide students with a "framed outline" that students can fill in and copy into their flip book. Depending on the needs of the students, this outline could be presented to the whole class, or provided for a small group of students. Either way, the teacher should model the use of the outline so that students have a clear idea as to how to use it correctly.

PLANNING POINT 1

The Flip book is described in *99 Ideas* book (p. 55).

PLANNING POINT 2

The Conga Line is similar to Inside-Outside Circle described in the *99 Ideas* book (p. 110). This is a fun, interactive wrap up for the students.

FIGURE 5.2 *Sun and Moon Phases Framed Summary Outline*

My favorite phase of the day is _______________. It is my favorite

because _______________. I like to see the _______________

in the sky. One activity I like to do during _______________ is

_______________. I also like to _______________.

Concluding Thoughts

We know that elementary teachers often like thematic lessons and cross-curricular opportunities. This unit could be extended and linked to other content areas easily. For example, there are many literature books that talk about sun, the moon, or the time of day. These books could be shared during reading or language arts time. Math concepts (e.g., telling time) and social studies concepts (e.g., map skills) could be connected to this unit as well.

We hope this unit gave you new ideas for incorporating meaningful activities and attention to language development in your science lessons. The young ELs are emergent readers and writers, so these lessons rely on visuals, discussion, and kinesthetic activities more than reading and writing texts. Helping students articulate their science knowledge using oral sentence frames was one goal of the unit, as was developing their vocabulary base. Notice, however, we did not avoid higher order tasks: The students were asked to classify and synthesize.

We encourage you to read Chapters 6–8 for additional science unit lessons. Even if the grades you currently teach focus on K–2, you will find interesting SIOP® science lessons and effective integration of the new techniques in these other sample units.

chapter 6

SIOP® Science Units and Lessons for Grades 3–5

Amy Ditton and Deborah Short

Introduction

This chapter, along with Chapters 5, 7, and 8, offers a complete five-day science unit including lesson plans and student handouts. In this chapter, SIOP® science educator Amy Ditton discusses in detail how she planned this unit for students in the upper elementary grades. She also focuses on the writing of one particular lesson, Day 5, and describes her decisions and selections when preparing it. (*Note*: If you have not read Chapter 4 about SIOP® science unit and lesson design for English learners, please read it for an overview of the planning process and a discussion of the lesson plan format used in this and other chapters.) As you read the unit here, you will see how several of the techniques and

activities from Chapters 2 and 3 are incorporated into the lessons, as well as how the activities in the unit lesson plans map onto the components and features of the SIOP® Model (found in Appendix A).

You will notice that there are Think-Alouds and Planning Points in many of the lessons. The Think-Alouds, identified with a thought bubble icon, are questions we asked ourselves and/or decisions we made as we undertook the SIOP® unit-planning process. The purpose of the Planning Points, identified with a flash drive icon, is to highlight or identify resources that support the implementation of the lesson.

Newton's First Law of Motion

Unit Overview, Grades 3–5

To plan this unit, I worked closely with Hope Austin-Phillips, a middle school science teacher who consistently and effectively implements the SIOP® Model in her classroom. I started with our state science and ELP standards, and then adapted lessons that I had seen Hope teach to meet the needs of the students in upper elementary classrooms. The collaboration we undertook for this unit was essential to its success and I had a lot of fun planning the lessons.

Clear Explanation of Academic Tasks

One thing that was important to accomplish throughout the unit was that students had a Clear Explanation of Academic Tasks (SIOP® Feature 11). I initially considered having the teacher model each experiment, but upon reflection realized it might eliminate some of the surprise from the mini-experiments. If the set-up were modeled however, then students could conduct the mini-experiments independently, provided that the instructions were clearly explained. That's the approach I decided on. Nevertheless, depending on your students, you may choose to model one or more of the mini-experiments.

Hands-on Experiences

On Day 1 of the unit, students immediately get the chance to participate in mini-experiments in order to experience Newton's First Law of Motion firsthand. While the lesson bustles with activity, the students are reading, writing, listening, and speaking throughout. Day 2 of the unit gives them a chance to record their observations from these mini-experiments in writing. Just writing independently in their science journals did not seem to provide enough organization and structure to connect students' concrete experiences to the abstract concept of Newton's First Law of Motion, so I developed the chart and sentence frames to support the task. This scaffolding helps students articulate the scientific language. On Day 3, the class again participates in hands-on, mini-experiments that make the abstract law of motion more concrete for them. They are introduced to new forces and conditions that have an effect on an object in motion or at rest.

Language Development

All the lessons provide multiple opportunities for language practice, through what I anticipate will be meaningful activities. But the language objectives for the days vary according to the topic and the students' academic literacy needs. On Day 2, as mentioned above, the focus is on writing, using sentence starters to bolster the students' entries. On Day 4, I went back and forth trying to decide which sentence starters to use. First, I thought about listing specific sentence starters for students, such as "One example of friction is _____," but I decided that after their concrete experiences with each type of force, students could determine the forces themselves and so I used a more generic sentence frame, "One observation I made was ______." To help students at lower proficiency levels, however, I could give them a frame with more specific information or I could pair them up with a more proficient student. With this lesson, I debated using Milling to Music again, but chose Pass the Note Card instead. The elementary students enjoy physical movement and it doesn't require a certain room set up for students to stand, pass, and read aloud. With the experiment centers already arranged, this technique would be an easy way to get students up and out of their seats.

Assessment

I like to wrap up each unit with an assessment that pulls the learning points together. On Day 5, the students conduct a final experiment in which they integrate their learning. Through the lab activity, they have a chance to practice and apply both the content and language objectives of the lessons.

Day 5 Lesson Design: Forces Every Day

The unit about Newton's First Law of Motion is fun for teachers and students alike because it includes so many mini-experiments. The fifth lesson in the unit is an opportunity for students to synthesize their learning from all of the mini-experiments and apply the language at a higher level than was required in previous lessons.

The lesson actually came about after I thought I had finished the unit. I was reviewing the five original lessons, and realized that one lesson in particular was really more "filler" than a lesson that actually extended student learning. (That lesson, since removed, had students reflect on the different forces and decide which force they found most interesting and why.) I noticed instead that the unit never summarized or "wrapped up" the key information; in brief, it lacked a final assessment. Each individual lesson was assessed, but student learning of the unit as a whole was not.

The unit needed to include an opportunity for students to synthesize and summarize their learning. If I provided an opportunity for students to focus on writing and using academic language without competing for time to do a hands-on experiment on same day, I could determine better if students had internalized both the language and content goals. I wrote the language objective for Day 5 then so students would summarize their learning in a paragraph. I included graphic organizers as scaffolds so students would be able to write a complete paragraph even if they had lower levels of English proficiency.

I decided that the final lesson should give students opportunities to think at higher levels, namely about how Newton's first law of motion affects our everyday life. While planning, I listed many examples that the teacher could share with the class, but came to realize that the teacher would still be "giving students the answer." Students would be pressed to think at higher levels if they heard only a few examples and then had to generate their own ideas. This notion became the basis for the Day 5 content objective.

The lesson begins with students participating in a Think-Pair-Share activity to discuss their favorite mini-experiment. Reviewing the mini-experiments is important because it gives students a chance to make connections to prior learning as well as to their background experiences. I used Think-Pair-Share because it is quick and I wanted students to spend the majority of their time synthesizing and writing.

I wanted students to develop their own ideas about how Newton's First Law of Motion affects our everyday life. Because it is a difficult concept, even after the mini-experiments, it is important for the teacher to share some examples as a starting point for the students. So I suggested teachers use examples like those on the Examples Chart (e.g., a soccer ball at rest and then kicked in the air). Depending on the language levels of the students, the teacher may need to incorporate visuals to go along with the examples that he/she shares.

I then paired students to brainstorm ideas of how Newton's first law affects our everyday lives. Some students may have a hard time thinking of an example, so pairing students is another way to support their comprehension. I wanted their explanations to include specifics, but students, particularly English learners, may have a hard time identifying and explaining everything that is happening without explicit scaffolding. So, I included a chart with key vocabulary to help students organize their thoughts. This graphic organizer also serves as a feedback tool, because if some students can't identify the different elements, they may realize that their idea is not an example of Newton's First Law of Motion. As with other tasks throughout the lesson, the teacher needs to model use of the Everyday Forces Chart so that students know exactly what is expected. Students continue to work in pairs as they fill out their Everyday Forces Charts.

Up until this point, most of the activities in this unit have been collaborative and hands on. For the writing activity, I wanted students to work independently. This would allow the teacher a more accurate assessment of student comprehension and academic language use. The way I designed the chart, students have support in identifying the different elements, but for the student to actually write about them, the concepts have to be internalized. Although each student is supported with sentence starters, if he or she does not understand the concepts, the final writing piece won't make sense.

For a "wrap up" activity, I chose Web of Information with the use of Outcome Sentences for a few different reasons. First, every student is responsible for sharing his/her learning with the whole class. The outcome sentences also allow students to share questions or thoughts that extend what they have learned. I think using the string and having something in students' hands in the Web of Information makes the activity more memorable than a quick Think-Pair-Share. The activity allows the class to formally wrap up the unit as a whole.

3–5 Unit

Now you will have a chance to read the unit on Newton's First Law of Motion. While you read these lessons, you can assess how well Amy Ditton met her unit and lesson design goals for the students. Some student worksheets are listed as BLMs (Blackline Masters) and are located in Appendix C.

SIOP® LESSON PLAN: *Grades 3–5 Unit: Newton's First Law of Motion, Day 1: First Law of Motion Experiments*

Developed by Amy Ditton

Key: SW = Students will **TW** = Teacher will **HOTS** = Higher Order Thinking Skills

Standard: Identify the conditions under which an object will continue in its state of motion (Newton's First Law of Motion).

Key Vocabulary:
object, rest, motion, force, gravity, friction

HOTS: Students summarize definitions of new vocabulary words and Newton's First Law of Motion in their own words.

Visuals/Resources/Supplementary Materials:
Vocabulary words with graphics (See Think-Aloud 1.)

Posted chart paper with Newton's First Law of Motion written on it

Sticky notes

Science journals

Digital camera

Note cards

Pennies

Beakers

Day 1 mini-experiment procedures sheet (Figure 6.1)

Mini-experiment observation notes (BLM #3)

Connections to Prior Knowledge/Provide Background Information:
- TW introduce content and language objectives.
- TW introduce key vocabulary using total physical response movements and illustrations (words with graphics chart).
- SW add vocabulary words to their science journals. Vocabulary entries will include teacher-given definition, definition in students' own words, and an illustration.

Objectives:	*Meaningful Activities:*	*Review/Assessment:*
Content Objective: SW conduct mini-experiments and make observations **Language Objectives:** SW paraphrase Newton's First Law of Motion in writing. SW report observations using complete sentences in the past tense.	• TW introduce Newton's First Law of Motion: An object at rest stays at rest and an object in motion stays in motion, unless a force acts upon it. • SW write Newton's First Law of Motion in their science journals. • TW explain she wants students to paraphrase this law. She gives an example: "It is against the law for teenagers to drink and then operate a car," which can be paraphrased as "Don't drink and drive." • SW think-pair-write-share, putting Newton's First Law of Motion in their own words on sticky notes. • SW post their sticky notes at the front of the room on the chart paper that displays the Law. • TW distribute the mini-procedures experiment and observation sheets (Figure 6.1 and BLM #3.) S/he will facilitate Mini-Experiment #1 (Running) with small groups and will model taking notes on the experiment.	• TW informally assess student responses in think-pair-write-share. • TW read written responses.

• As groups take turns running, other groups take notes on what they observe on the observation sheet. • TW model the set-up for Mini-experiments #2 and #3 (Flick the Note card and Pile of Pennies) and will refer students to the procedures posters for each experiment. • SW work in small groups to conduct Mini-Experiments #2 and #3 (students can take turns conducting each experiment) and will make observations about Mini-Experiments #2 and #3. • TW model using notes to write an observation in a complete sentence in the past tense, using a sentence starter: One observation I made during the experiment was ____. (See Think-Aloud 2.)	• TW circulate as students work in groups. • TW ask questions to each group as they work.
• SW use their notes to write one complete sentence about each experiment on their observation sheet.	• TW collect written responses.

Wrap Up:

- SW share their written sentences in a "Milling to Music" activity (see procedure in Chapter 3).
- TW refer students back to the content and language objectives to ask if they were met, and if so, how they were met.

(Template developed by Melissa Castillo and Nicole Teyechea. Used with permission.)

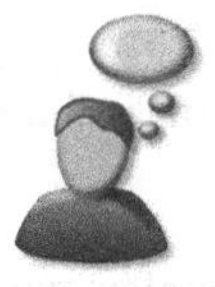

THINK-ALOUD 1

At the start of this unit, I prepare a vocabulary chart with pictures or graphics to illustrate the key words. I use it to introduce the words and keep it posted in the room throughout the unit. It is helpful for words that are hard to visualize, like gravity. So my illustration is an apple falling with an arrow indicating the downward movement.

THINK-ALOUD 2

I have found that it is very helpful for English learners and some other students to have a model and a sentence starter to use so they practice appropriate scientific language and discourse style. More advanced students may write observations on their own, although they also need to use complete sentences and past tense verbs.

FIGURE 6.1 *Day 1 Mini-Experiments Procedures*

Running Experiment

1. Students line up outside, as if preparing to race, on the basketball court, playground, or other available surface.
2. When the teacher blows the whistle, students run as fast as they can.
3. After 3–4 seconds, the teacher blows the whistle again and students stop as quickly as they can.
4. A teacher, parent volunteer, classroom assistant, or student should take digital photos as students are trying to stop.

Penny Experiment

1. Place a beaker in front of you on your table or desk.
2. Place a note card on top of the beaker.
3. Place a penny on top of the note card.
4. Flick the note card off of the beaker.
5. Notice what happens to the penny! (The penny should fall into the beaker demonstrating the force of gravity.)

Pile of Pennies

1. Stack 5 pennies, one on top of another.
2. Place one penny about 1–2 inches away from the stack of pennies.
3. Flick the single penny toward the stack of pennies. (The only penny in the stack to move should be the bottom penny.)

SIOP® LESSON PLAN: *Grades 3–5 Unit: Newton's First Law of Motion, Day 2: Explaining the Mini-Experiments*

Developed by Amy Ditton

Key: SW = Students will **TW** = Teacher will **HOTS** = Higher Order Thinking Skills

Standard: Identify the conditions under which an object will continue in its state of motion (Newton's First Law of Motion).

Key Vocabulary:
object, rest, motion, force, gravity, friction

HOTS: Students analyze Mini-Experiments and synthesize their observations with Newton's First Law of Motion.

Visuals/Resources/Supplementary Materials:
Vocabulary words with graphics posted in classroom

Chart paper with Newton's First Law of Motion written on it, posted in room

Science journals

Mini-Experiments Summary Chart (Figure 6.2)

Ball of yarn or string

Connections to Prior Knowledge/Provide Background Information:
- TW introduce content and language objectives.
- TW ask students to perform "Oh Yesterday!" (see procedures in Chapter 2) to relate at least 1 observation for each of the mini-experiments from the previous day. Students may refer back to their notes or science journals as they recall their observations.
- TW show the digital photos from the day before (if they were taken).

Objectives:	*Meaningful Activities:*	*Review/Assessment:*
Content Objective: SW apply Newton's Law of Motion to their observations from the mini-experiments.	• TW use the Mini-Experiments Summary Chart (Figure 6.2) to model applying Newton's First Law of Motion to one observation about a mini-experiment.	• TW circulate the room asking groups questions and providing feedback as students work.

Language Objective: SW use key vocabulary and sentence frames to explain their observations from the mini-experiments. ______ stayed at rest because . . . ______ was in motion. ______ was the force that acted on . . .	• SW work with the same small groups from the previous lesson to apply Newton's First Law of Motion to their observations of the three mini-experiments. Each student will record his/her responses in the boxes of the top half of the summary chart. • SW share their responses in a Numbered Heads Together activity. • TW next model taking a description from the summary chart boxes and explaining the description using key vocabulary and sentence starters. Students will complete the bottom half of the summary chart sheet.	• TW assess student responses in the Numbered Heads Together activity. • TW collect summary charts to read student responses.

Wrap Up:

- SW share their explanations using the "Web of Information" activity (see procedures in Chapter 3). The class will sit in a circle and as each student shares an explanation, he or she hangs on to a piece of yarn or string before tossing the ball to the next speaker.
- TW refer students back to the content and language objectives and will ask students if they were met, and if so, how.

(Template developed by Melissa Castillo and Nicole Teyechea. Used with permission.)

FIGURE 6.2 *Mini-Experiments Summary Chart for Day 2*

An object at rest stays at rest
AND
an object in motion stays in motion
UNLESS
a force acts upon it.

Mini-Experiment	Running	Flick the Note card	Pile of Pennies
Observation	I fell when I stopped		
Object at Rest			
Object in Motion	Me		
Force	Friction with my feet when I tried to stop.		

Sentence Frames:

- ______ **stayed at rest until . . .**
- ______ **was in motion until . . .**
- ______**was the force that acted on . . .**

Running Explanations:
I was in motion until I used my feet to try to stop.
Friction was the force that acted on me.

Flick the Note Card Explanations:

Pile of Pennies Explanations:

SIOP® LESSON PLAN: *Grades 3–5 Unit: Newton's First Law of Motion, Day 3: Forces Are Fun* Developed by Amy Ditton

Key: SW = Students will **TW** = Teacher will **HOTS** = Higher Order Thinking Skills

Standard: Identify the conditions under which an object will continue in its state of motion (Newton's First Law of Motion).

Key Vocabulary:
object, rest, motion, force, gravity, friction, centripetal force, magnetic force, buoyant force

HOTS: Students have to summarize definitions of new vocabulary words in their own words.

Visuals/Resources/Supplementary Materials:

Posted vocabulary words with graphics

Posted chart paper with Newton's First Law of Motion written on it

Science journals

Bucket

Magnets and small metal objects

Beakers with water and small objects (e.g., paper clips, grapes, styrofoam peanuts)

Day 3 Mini-Experiment Procedures (Figure 6.3) (See Planning Point 1.)

Forces Are Fun Observation Page (BLM #4)

Forces Are Fun Summary Page (BLM #5)

Connections to Prior Knowledge/Provide Background Information:

- TW introduce content and language objectives.
- SW refer back to their completed Mini-Experiments Summary Chart (Figure 6.2) from the previous lesson and will respond to the following questions in their science journals:

  What was the most interesting observation you made yesterday?
  What ideas do you have for another mini-experiment demonstrating Newton's First Law of Motion?
- SW share their responses with a partner. TW randomly call on students to share responses with the whole class.
- TW introduce new vocabulary words (*friction, centripetal force, magnetic force, buoyant force*) using total physical response and graphic illustrations.
- SW add vocabulary words to their science journals. Vocabulary entries will include teacher-given definition, definition in students' own words, and an illustration.

Objectives:	*Meaningful Activities:*	*Review/Assessment:*
Content Objective: SW conduct five mini-experiments relating to force and make two observations about each. **Language Objective:** Students will use the following sentence starters to reflect on their observations in writing: "One interesting thing I noticed was . . ." "_______was an example of _______."	• TW demonstrate Mini-Experiment #1 (Bucket of Water). (See Figure 6.3.) • SW write two observations about Mini-Experiment #1 on the observation page (BLM #4). • TW model using a sentence starter to explain an observation from "Bucket of Water" and SW record on the observation page. • TW explain Mini-Experiments #2–#5 and will refer students to the procedures at the lab stations. (These mini-experiments are Magnets and Metal, Forces in Water, Paper Drop, and Hot Hands.) • SW work in small groups to conduct Mini-Experiments #2–#5. They move to stations around the classroom.	• TW circulate as students work in groups.

• SW work in small groups to write observations about Mini-Experiments #2–#5 on the observation page. • After the mini-experiments are done, TW model using a sentence starter to reflect on an observation from "Bucket of Water" and SW follow on the summary page (BLM #5). • SW work in small groups to complete the Forces Are Fun Chart and write summaries about Mini-Experiments #2–#5 on the summary page (BLM #5).	• TW ask questions to each group as they work.

Wrap Up:

- SW share their written sentences in a "Milling to Music" activity (procedures in Chapter 3). TW then collect written responses.
- TW refer students back to the content and language objectives to ask if they were met, and if so, how they were met.

(Template developed by Melissa Castillo and Nicole Teyechea. Used with permission.)

PLANNING POINT 1

It is best to set up the mini-experiments #2–#5 in lab stations around the room. I prepare the stations with the steps and supplies before the class begins. I cut the Mini-Experiment Procedures sheet (Figure 6.3) apart for each experiment and display each experiment's steps at the appropriate station.

FIGURE 6.3 *Day 3 Mini-Experiment Procedures*

1. **Bucket of Water**
   1. Fill a bucket with water.
   2. Hold the bucket by the handle with your arm straight out in front of you.
   3. With the bucket held away from you, spin in circles. The water should stay in the bucket.
2. **Magnets and Metal**
   1. Provide students with magnets and metallic objects.
   2. Have students play with the magnets and the objects for a few minutes.
3. **Forces in Water**
   1. Provide students with beakers of water and small objects. (Be sure that at least one of the objects will float. Styrofoam peanuts, for example, will work.)
   2. Have students drop the objects into the water to see what happens.
4. **Paper Drop**
   1. Provide students with small pieces of paper.
   2. Have students drop the pieces of paper to the ground from different heights (e.g., above their head, waist level).
5. **Hot Hands**
   1. Have students place their hands together, fingertips and palms touching.
   2. Instruct students to rub their hands together slowly and then quickly.

SIOP® LESSON PLAN: *Grades 3–5, Unit: Newton's First Law of Motion, Day 4: Raisin Experiment* Developed by Amy Ditton

Key: SW = Students will **TW** = Teacher will **HOTS** = Higher Order Thinking Skills

Standard: Identify the conditions under which an object will continue in its state of motion (Newton's First Law of Motion).

Key Vocabulary:
object, rest, motion, force, gravity, friction, centripetal force, magnetic force, buoyant force

HOTS: Students synthesize their knowledge of an experiment, forces, and Newton's First Law of Motion.

Visuals/Resources/Supplementary Materials:

Posted vocabulary words with graphics

Posted chart paper with Newton's First Law of Motion written on it

Overhead projector

600 ml beakers (one per student group)

Baking soda

Vinegar

Raisins

Sticky notes

Procedures for Raisin Experiment (Figure 6.4)

Mysterious Raisins Lab Report (BLM #6)

Connections to Prior Knowledge/Provide Background Information:
TW show a slide show of students participating in the Running Mini-Experiment, demonstrating Newton's First Law of Motion and refer students to the observation and summary pages completed in the previous lesson. SW participate in a "Graffiti Write" to recall everything they remember about Newton's First Law of Motion and the vocabulary words. SW work in groups of 3 or 4, each with a different color marker, and one group sheet of chart paper. Using their markers, each student writes his/her name on the back of the paper. Then, at the teacher's signal, students write everything they remember for 1–2 minutes on the front of the chart paper. Students are allowed to discuss their responses only after time is up and everyone is finished writing. (See Think-Aloud 1.)

Objectives:	*Meaningful Activities:*	*Review/Assessment:*
Content Objectives: SW make 6–8 observations during the raisins in the beaker experiment.	• TW model the procedures for the raisin experiment (see Figure 6.4) and how to state the language objective by sharing two observations using the sentence starter. • SW work in small groups (3–4) to complete the raisin experiment.	
SW apply the concepts of gravity, buoyant force, and Newton's First Law of Motion to explain what happened in the beaker.	• As students observe the experiment, they will record 6–8 observations on the "Mysterious Raisins" Lab Report handout (BLM # 6). • TW randomly call on students to share one observation from their notes.	• TW assess the students' written observations. TW also informally assess the content objectives by circulating among the groups and asking questions.
Language Objectives: SW discuss observations and the application of Newton's First Law of Motion with a partner.	• TW then use one of the shared observations to model the use of the terms *buoyant force* and *gravity* to describe what was happening in the beaker. Students will use the terms *buoyant force* and *gravity* for two of their observations and will write their descriptions on the Mysterious Raisins Lab Report handout. (See Think-Aloud 2.)	• TW read student explanations.

SW use key content vocabulary words and the given sentence starter to write about their observations: One *observation* I made was ________	• TW take another shared observation and explain the observation using Newton's First Law of Motion. TW write the example on the overhead. • TW pick a third shared observation, discuss it, and write its explanation on the overhead. • SW work with a partner to use Newton's First Law of Motion to explain one observation and write the explanation on the Mysterious Raisins Lab Report handout. • TW ask students to share their explanations with the whole class. • SW choose three more of their observations and write explanations using Newton's First Law of Motion for each observation. They will write the explanations on the Mysterious Raisins Lab Report handout and then transfer three of them to three sticky notes. SW place two notes on the chart on the board and keep one for the final activity.	• TW review student explanations.

Wrap Up:
- SW share their explanations in a Pass the Note Card activity (see procedure in Chapter 3). SW pass their sticky note around to others in a circle while music plays. When the music stops, TW randomly select people to read what is on the sticky note they hold at that time and discuss.
- TW refer students back to the content and language objectives to ask if they were met, and if so, how they were met.

(Template developed by Melissa Castillo and Nicole Teyechea. Used with permission.)

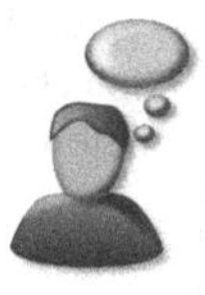

THINK-ALOUD 1

This is a fun way to review prior knowledge. Students work in groups, but don't discuss at first. I can assess individual student comprehension by matching the color of the marker to the child's name written in the same color on the back of the chart paper.

THINK-ALOUD 2

I may pair up my students who have a low proficiency level with students who have stronger academic English skills to help them express their observations in terms of Newton's first law.

FIGURE 6.4 ***Procedures for the Mysterious Raisins Experiment***

Follow these procedures for the Mysterious Raisins Experiment:

1. Put 100 ml of water in your beaker.
2. Add one spoonful of baking soda.
3. Add 5 raisins.
4. Add 15 ml vinegar.
5. Record your observations for 2–3 minutes.
6. After 2–3 minutes, add another 15 ml of vinegar to your beaker.
7. Record your observations for 2–3 more minutes. You should have at least 8 observations in total now.
8. Clean Up! Rinse the beaker with water. Throw the raisins in the wastebasket.

SIOP® LESSON PLAN: *Grades 3–5 Unit: Newton's First Law of Motion, Day 5: Forces Every Day* Developed by Amy Ditton

Key: SW = Students will **TW** = Teacher will **HOTS** = Higher Order Thinking Skills

Standard: Identify the conditions under which an object will continue in its state of motion (Newton's First Law of Motion).

Key Vocabulary:
object, rest, motion, force, gravity, friction, buoyant force, centripetal force

HOTS: Students will synthesize what they know about Newton's First Law of Motion.

Visuals/Resources/Supplementary Materials:
Everyday Forces Chart (BLM #7)

Overhead transparency of Examples Chart of Everyday Forces (Figure 6.5)

Writer's Checklist (BLM #8)

Connections to Prior Knowledge/Provide Background Information:
- TW introduce content and language objectives.
- Using think-pair-share, SW explain what their favorite experiment in the unit was and why they chose it.

Objectives:	*Meaningful Activities:*	*Review/Assessment:*
Content Objective: SW identify examples of Newton's First Law of Motion in everyday life. **Language Objective:** SW write a paragraph explaining how Newton's First Law of Motion affects everyday life.	• TW distribute an Everyday Forces Chart to each student (BLM #7). TW share examples of Newton's First Law of Motion in everyday life using the top half of the transparency to Examples Chart of Everyday Forces (see Figure 6.5). • TW model how to complete the Everyday Forces Chart using the sample on the transparency of the Examples Chart (bottom half) to describe what happens in everyday life that reflects Newton's First Law of Motion. • SW work with a partner to think of an everyday example of Newton's First Law of Motion. • SW work with a partner to fill out an example on their own Everyday Forces Chart. (Each student fills out his/her own chart.) • TW model using the chart and sentence starters on the Everyday Forces Chart to write a paragraph explaining how Newton's First Law of Motion affects everyday life. Teacher may show the paragraph on the bottom of the Examples Chart transparency as a model for students. • SW work individually to use their charts and provided sentence starters to write a paragraph explaining how Newton's First Law of Motion affects everyday life. • SW use the Writer's Checklist (BLM #8) to evaluate their own work.	• TW circulate the room to monitor comprehension and language use. • TW read students' completed Examples Charts to check comprehension and language use. • TW use the writer's checklist to evaluate student written paragraphs.

Wrap Up:

- SW use outcome sentences to share their learning in a Web of Information activity (see procedures in Chapter 3).

  I wonder . . .
  I discovered . . .
  I still want to know . . .
  I learned . . .
  I still don't understand . . .
  I still have a question about . . .
  I will ask a friend about . . .

(Template developed by Melissa Castillo and Nicole Teyechea. Used with permission.)

FIGURE 6.5 *Examples Chart of Everyday Forces*

Examples of Newton's First Law of Motion

- Someone is riding on a skateboard and he or she hits a curb. What happens?
- An astronaut is on a space walk and throws an object upward. What happens?
- There is a soccer ball that was just kicked into the air. It travels about 15 feet and then falls back to the ground. What happened?
- There are two people sitting at opposite ends of a long table. The first person tries to slide a book across the table to the other person, but it stops midway. What happened?

real life object	at rest or in motion	changes to motion or rest	force acting on the object	type of force (friction, gravity, centripetal, magnetic, buoyant)	What happened?
soccer ball	in motion	becomes at rest	air	friction and gravity	• air/ball friction slows the ball down • gravity brings the ball back to the ground • the ball won't move again until a force acts upon it

A soccer ball traveling through the air and falling to the ground is an example of Newton's First Law of Motion. When the soccer ball is in motion, it becomes at rest. Friction between the ball and the air affects the ball and slows it down. Gravity affects the soccer ball and brings it back to the ground. Due to Newton's First Law, the soccer ball will then remain at rest until another force, like someone kicking it, acts upon it.

Concluding Thoughts

When students reach the upper elementary grades, they are ready for more abstract science concepts but still need concrete experiences to understand them well. The numerous mini-experiments in this unit provide that real-life support as students build their comprehension of Newton's first law of motion. The difficulty and duration of the experiments build over the week as well. The lessons involve a good deal of scientific writing as students record observations and summarize information. The final writing task requires them to synthesize their knowledge, which is a useful process for the types of lab reports they will be writing at the middle school level.

We hope this unit gave you new ideas for incorporating meaningful activities and attention to language development in your science lessons. We encourage you to read Chapters 5, 7, and 8 for additional unit lessons. Even if the grades you currently teach focus on grades 3–5, you will find interesting SIOP® science lessons and effective integration of the new techniques in these other sample units.

chapter 7

SIOP® Science Units and Lessons for Grades 6–8

Hope Austin-Phillips and Deborah Short

Introduction

You will find in this chapter a complete five-day science unit, including lesson plans and handouts for middle school students. As in Chapters 5, 6, and 8, the chapter opens with a detailed discussion of the thinking process that went into the unit design by one of our SIOP® science educators, Hope Austin-Phillips. She offers her insights for planning this middle school SIOP® science unit on the rock cycle and also describes how she decided on activities and resources for the lesson plan for Day 2. (*Note*: If you have not read Chapter 4 about SIOP® science unit and lesson design for English learners, please read it for an overview of the planning process and a discussion of the lesson plan format used in

this and other chapters.) As you read this unit, you will see how several of the techniques from Chapters 2 and 3 are incorporated into the lessons, as well as how the activities in the unit lesson plans map onto the components and features of the SIOP® Model (found in Appendix A).

You will notice that there are teacher Think-Alouds and Planning Points in many of the lessons. The Think-Alouds, identified with a thought bubble icon, are questions we asked ourselves and/or decisions we made as we undertook the SIOP® unit-planning process. The purpose of the Planning Points, identified with a flash drive icon, is to highlight or identify resources that support the implementation of the lesson.

Science Rocks!

Unit Overview, Grades 6–8

Teaching the rock cycle to middle school students is very enjoyable. They tend to become very excited about rocks and love the hands-on aspects of these lessons. For this unit, I present the students with a packet of worksheets and graphic organizers to help them keep their notes together.

Vocabulary

One of the most important aspects of teaching science to English learners (or any student for that matter) is making sure that they have access to sophisticated science vocabulary words. It is essential to say the words repeatedly. It is also imperative to have the words in writing and with graphics for the students to view. When the students have repeated exposure to a word by hearing it, seeing it in writing, seeing it in a graphic representation, and using it, they are able to incorporate it into their own expressive vocabulary. Knowing these terms and the related concepts helps to make the content topics more comprehensible.

When introducing sedimentary rocks, for instance, every time the term *sedimentary rock* is said, I hold up a word card with a graphic representation. I walk around the room holding the word so students can see it at all times. After the lesson, I post it and other key terms from the lesson on the bulletin board so students have access to the written word and an illustration during the entire unit.

An example of how this can be effective is that a student might say or write, "The rocks have little rocks in them." I probe, "What kind of rocks?" If the student does not remember the vocabulary word, I might say, "What kind of rocks have we been studying? It is one of our vocabulary words. It is posted on the board." With these clues, the student should be able to locate the word on the word wall and sound it out. If necessary, I may support him or her with the pronunciation.

Hands-On Experiences

While teaching the rock cycle to English learners, I find it is very important to give them hands-on experiences to build background knowledge. In this unit, the students have multiple chances to observe different types of rocks closely before reading about them. Often students make insightful observations about a rock, which I, or they, can refer back to in a later lesson, perhaps during a reading activity or a lab experiment.

For example, while my students are observing sedimentary rocks, most students will notice that some rocks are made up of "many little rocks." They will also observe that some rocks have fossils. In a later lesson, when the students read about sedimentary rocks, they learn about detrital sedimentary rocks. Detrital sedimentary rocks are made up of many tiny fragments of rocks cemented together. While I introduce the vocabulary word, *detrital* I will refer to the students' previous observations. I might even say, "Raise your hand if you noticed many tiny rocks in one of the sedimentary rocks." After the students raise their hands (many students will have noticed this), I might praise them and expand, "You are acting like real scientists! Scientists also have noticed the little rocks inside some sedimentary rocks and they called these *detrital sedimentary rocks.*"

Modeling, Guided Practice, and Independent Practice

I designed these lessons in order for students to have a lot of modeling and guided practice on Day 2 of the unit. By Day 3 then, the students will have already rehearsed the skills they need to work independently on the assignment. I monitor the students at this point and use positive and specific feedback to guide them. I typically state the feedback aloud so the entire class can benefit and incorporate suggestions into their work.

Further Investigation into the Rock Cycle

This unit is meant to be an introduction to rocks and the rock cycle that the students need as background information. It is important for them to learn the vocabulary and basic information first. As part of a more rigorous science curriculum, students will further investigate the Earth processes that are involved in the rock cycle in the subsequent unit. The Science Rocks packet can be saved and used in the next unit.

Once I have the unit outlined, I design each lesson carefully. Next, you can follow my thinking process as I wrote the Day 2 lesson for Science Rocks.

Day 2 Lesson Design: Sedimentary Rocks

Students really enjoy studying rocks, especially when I incorporate hands-on activities and scientific observations. So, before this Day 2 lesson, the students spend a day making observations about rocks. I give them a tray full of igneous, sedimentary, and metamorphic rocks. They use magnifying glasses to closely observe the rocks. Then, they categorize the rocks based on characteristics they have seen. I always do the lesson this way because then students come into class on the second day with background knowledge and personal experiences with rocks. Moreover, I often find that students have never taken the time to simply observe rocks before and such observations are very important.

On Day 2 of the rock cycle unit, the first thing I want students to do is tap into their background knowledge and experiences from Day 1. Milling to Music is one of my favorite activities to help students recall information from previous classes. The students hold their notes from Day 1 as they move around the classroom. The activity only takes five minutes, yet the students have a chance to talk to four or five different partners during this time.

After Milling to Music, I pull some "magic name sticks" (craft sticks with student names) from a cup to randomly choose students to share their observations. I found that when I started using the magic name sticks, students became more invested in each activity because they knew I might call on them at any time. I only use this technique *after* I have given the students a chance to talk with peers, however. This way, they have already had a chance to share with a partner or small group before they are expected to share with the class.

My goal for this lesson is for students to become familiar with sedimentary rocks. In order for them to really learn about sedimentary rocks, they need to have hands-on experiences. For this reason, I find the Alternate Materials technique essential for this lesson. The students examine a tray full of chemical, organic, and detrital sedimentary rocks. Some of these rocks might include breccia, conglomerate, limestone, shale, white and red sandstone, bituminous, and coquina. I always include at least one or two samples of rock with visible fossils.

It is essential that this hands-on inquiry activity be conducted before students begin to read about sedimentary rocks. I have found that if students make clear scientific observations about the rocks first, they will make better connections with the reading later. For example, in my experience, students always notice the fossils. This is a key fact: Sedimentary rocks are the type of rock where scientists find fossils. Students also always notice that conglomerate and breccia are composed of many rocky fragments. This is another key fact about sedimentary rocks. When the students begin the reading part of the lesson, they will already have this background knowledge to apply to their new learning.

After the students make their observations about sedimentary rocks, I encourage them to share with the class so everyone will hear a broad range of observations. After the students report on their observations, I introduce the vocabulary words and try to connect each word to the observations. For example, when I introduce the word *sediment*, I will ask students to find the samples of breccia and conglomerate and I will tell them that the little rock fragments inside of these two rocks are called *sediment*. I also have a tub of sand and gravel that I display when we discuss sediment.

Next, the students are ready to start the reading activity. I typically find a passage about sedimentary rocks from a textbook or the Internet. I will have made copies of the reading selection and I distribute them along with a note-taking graphic organizer so the students can underline or highlight the text as we read. I also display the selection either with an overhead projector or a document camera.

To begin, I model the activity first on my own. I read the first few sentences aloud and then think-aloud as I underline important information in the text. I next model for students how I put the information in my own words before I write it on the note-taking organizer. The students copy this on their organizers. After this, I read a few more sentences aloud and the students underline important information and put it into their own words. I call on students to share their ideas before going on. If the students seem to have the hang of it, I let them finish the reading and recording of the facts on their own. If not, I model some more or work with a small group of students who need extra support.

When I am working with English learners, I find that science can be intimidating for them because there are many big words. As a result of this lesson, I want students to use words like "sedimentary rocks" and articulate facts they find out about them, so that is the focus of my language objective. Students often have a hard time saying these words and I want them to be comfortable with them. So, when the students have completed their note-taking organizers, I ask them to pair-share with a partner. They take turns reading all of their facts using the vocabulary words. The result is that they have written the facts and said the facts aloud.

Finally, for review and assessment, the students participate in a game of Snowball! I choose Snowball! because it is a great activity to help students review facts they have learned. In this case, students choose one of the facts they have recorded to write on the Snowball! graphic organizer. As the activity proceeds, they have a chance to read at least two other facts and will write a total of three facts. At the end of Snowball!, I use the magic name sticks again to select students to share facts about sedimentary rocks.

After this, I review the content and language goals of the day and ask students if they feel we have met the goals. If time remains, I will call on a few students to state something interesting they learned.

6–8 Unit

The full unit that Hope Austin-Phillips described is presented next. Notice how she incorporates discovery learning and science language development in these interactive lessons. Some student worksheets are listed as BLMs (Blackline Masters) and are located in Appendix C.

SIOP® LESSON PLAN: *Grades 6–8 Unit: Science Rocks, Day 1: Classifying Rocks Based on Observable Characteristics*

Developed by Hope Austin-Phillips

Key: SW = Students will **TW** = Teacher will **HOTS** = Higher Order Thinking Skills

Content Standard(s): Apply the following scientific processes to other problem solving or decision making situations: observing and classifying.

Distinguish the components and characteristics of the rock cycle for the following types of rocks: igneous, metamorphic, sedimentary.

Key Vocabulary:
characteristic, classify, category, observation

HOTS: Justifying; classifying.

Visuals/Resources/Supplementary Materials:
Samples of igneous, sedimentary, and metamorphic rocks (as in the Alternate Materials technique in Chapter 2)

Magnifying glasses

Sticky notes

Rock Cycle Vocabulary graphics (BLM #9)

Science Rocks! Unit Packet (page 1) (BLM #10)

Connections to Prior Knowledge/Building Background:
- TW ask students what they already know about rocks. SW think-pair-share what they know.
- TW ask students what questions they have about rocks. SW think-pair-share their questions and teacher records them on the board or chart paper.
- TW tell students that over the course of the next few weeks, they will be learning about rocks and the rock cycle. TW review the content and language objectives with the class.
- TW introduce vocabulary words using the rock cycle vocabulary graphics (see BLM #9).
- TW distribute the Science Rocks! Unit Packet (BLM #10).

(continued)

Objectives:	*Meaningful Activities:*	*Review/Assessment:*
Content Objectives: SW make detailed scientific observations about characteristics of rocks. SW classify rocks into categories based on their characteristics. **Language Objectives:** SW use vocabulary and sentence frames to explain their rationale for the selected categories: These rocks are classified in this category because . . . SW write eight observations about rocks.	• TW tell students that they will be making observations about rocks. • TW think-aloud to demonstrate to students how to make detailed, scientific observations and use adjectives related to shape, texture, color, size, and so on.	
	• TW model for students how to write an observation using a random object such as a water bottle. SW write-pair-share one observation about the rocks in the chart on page 1 of the Science Rocks! packet (BLM #10). TW give specific and positive feedback to students. SW write seven more observations about the rocks in the chart.	• SW write one of their observations on a sticky note, read it to a partner, and share it with the class.
	• TW review the words *category* and *classify*. TW tell students they are going to be creating a sorting tree to categorize their rocks using the characteristics.	
	• TW model a **sorting tree** for the class using different types of shoes. She puts a pile of shoes on a table and separates them into two groups, based on a feature she explains (e.g., brown vs. black color). Then she breaks these piles into two based on another feature (e.g., sandals vs. closed shoes), so she has four piles in total. This process can continue according to additional features.	
	• SW create a sorting tree based on the characteristics of their rocks. SW organize their piles on the desk. TW distribute sticky notes to groups, one per pile. SW write complete sentences explaining their rationale for classifying rocks into each category on the sticky notes. TW first model one sentence for the students using the shoe sorting tree. SW think-pair-share one sentence. TW give specific feedback to students. SW then write more sentences for each category in their sorting tree.	• TW monitor and assess students' sentences.

Wrap-Up: TW review posted content and language objectives. SW complete a ticket out of the classroom by writing one sentence using the language objective's sentence frame prompt on an index card or piece of paper and handing it to the teacher as they leave.

(Template developed by Melissa Castillo & Nicole Teyechea. Used with permission.)

SIOP® LESSON PLAN: *Grades 6–8 Unit: Science Rocks, Day 2: Sedimentary Rocks* Developed by Hope Austin-Phillips

Key: **SW** = Students will **TW** = Teacher will **HOTS** = Higher Order Thinking Skills

Content Standard(s): Distinguish the components and characteristics of the rock cycle for the following types of rocks: igneous, metamorphic, and sedimentary.

Key Vocabulary:
sedimentary, observation, detrital, chemical, organic, sediment, fossil

HOTS: Identifying important facts in the text; composing contextualized sentences; and summarizing facts in own words.

Visuals/Resources/Supplementary Materials:

Trays of mixed sedimentary rocks, such as breccia, conglomerate, limestone, shale, sandstone (white and red), bituminous, coquina, and rocks with visible fossils

Magnifying glasses

Science Rocks! unit packet (pages 2 and 3, BLM #10)

Vocabulary with graphics chart: sedimentary rocks, detrital, chemical, organic, fossil

Small container of sand and dust with a sign that says, "sediment"

Reading selection on rock cycle (from textbook) (See Planning Point 1)

Overhead projector with transparency (or another means of projection) of reading selection about sedimentary rocks

Connections to Prior Knowledge/Building Background:

- SW participate in a Milling to Music activity to review observations from yesterday (see procedures in Chapter 3). SW hold their written observations from Day 1 with them during the activity. SW share with 3–4 partners during the activity. TW pull magic names sticks (craft sticks with student names) (see Think-Aloud 1) from a cup to call on students randomly to share with the class. TW use specific, positive, and informative feedback to respond to each observation. TW emphasize the importance of including details in each observation.
- TW tell students that today they are going to be continuing their study of rocks. Today they will be looking at just one type of rock called *sedimentary rocks*. TW discuss posted content and language objectives.

Objectives:	*Meaningful Activities:*	*Review/Assessment:*
Content Objectives: SW make five detailed observations about sedimentary rocks. SW identify key facts about sedimentary rocks. **Language Objectives:** SW both write and orally share five important facts in their own words with a partner One fact about sedimentary rocks is . . . SW read for specific information.	• TW remind students that they have been practicing making observations. Today they are going to use their skills to make observations of sedimentary rocks. • TW hand out trays of sedimentary rocks with magnifying glasses. TW choose a rock randomly and model making an observation to remind students about writing detailed scientific observations. Student groups will record observations about sedimentary rocks on page 2 of the unit packet (BLM #10). • SW pair-share their observations. TW use magic name sticks to call on students to share observations. TW use the student observations to introduce important facts about sedimentary rocks. For example, students might notice the fossils in the rocks.	• TW check for details during the sharing of the think-pair-share.

(continued)

TW say "So you already know an important fact about sedimentary rocks. Now we are going to read about sedimentary rocks to find out what else scientists know about sedimentary rocks."	
• TW introduce the vocabulary words *detrital, chemical, organic,* and *sediment* by connecting terms to the students' observations and real objects (e.g., tub of sand and gravel). SW record definitions in their notebooks.	
• TW display the text selection about the rock cycle on a transparency and then read aloud part of the selection. TW think-aloud while underlining important information in the text and demonstrating how to put the facts into her/his own words. SW practice by underlining one fact they find in the reading. TW give specific feeback to students as they underline their first facts. SW finish reading the selection and will underline important facts. (See Planning Points 1 and 2.)	• TW monitor what text the students are underlining and give feedback.
• SW write the important facts from the reading on page 3 of their unit packet (BLM #10). SW pair-share with a partner: Student 1 will read a fact and then Student 2 will read a fact. They will continue to share until both students have read five facts.	• TW circulate and review the student notes in the packets.
• The class will do a Snowball! activity next. (See procedures for Snowball! in Chapter 3.) SW write one fact on the Snowball! graphic organizer using the language objective sentence starter. SW toss their Snowball! and get a new snowball. SW read the fact about sedimentary rocks and then write a different fact about sedimentary rocks. SW get a final snowball and will write one more fact about sedimentary rocks. SW find their original snowball and will read all three facts.	• TW check on student comprehension during the reporting out of Snowball facts.
• TW call on students using magic name sticks. SW read one of the three sentences to the class.	

Wrap-Up: TW review content and language goals for the day. TW tell students to think-pair-share the most interesting thing they learned today. TW call on student volunteers to share with the class.

(Template developed by Melissa Castillo & Nicole Teyechea. Used with permission.)

THINK-ALOUD 1

I write one student name per craft stick and keep them in a cup or jar. When I want to call on students randomly, I pull out a name. Sometimes I don't put the sticks back in until the activity is over, or sometimes until all sticks have been pulled. Depending on the question, if I pull a name but the student doesn't have his or her hand raised, I won't call on that student.

PLANNING POINT 1

You may want to look for text on the Internet to present the information and model the note-taking activity. Some websites have excellent illustrations.

PLANNING POINT 2

If you expect that some students will need your support to complete the note-taking organizer (e.g., beginning level ELs), arrange the desks or tables in one corner of the room where they can come and work with you. The other students may continue on their own.

SIOP® LESSON PLAN: *Grades 6–8 Unit: Science Rocks, Day 3: Igneous and Metamorphic Rocks*

Developed by Hope Austin-Phillips

Key: SW = Students will **TW** = Teacher will **HOTS** = Higher Order Thinking Skills

Content Standard(s): Distinguish the components and characteristics of the rock cycle for the following types of rocks: igneous, metamorphic, and sedimentary.

Key Vocabulary:
igneous, magma, lava, intrusive, extrusive, metamorphic, pressure, foliated, non-foliated

HOTS: Identifying facts in the text; composing contextualized sentences.

Visuals/Resources/Supplementary Materials:

Trays of mixed igneous rocks and trays of mixed metamorphic rocks

Magnifying glasses, sticky notes

Vocabulary with graphics: igneous, magma, lava, intrusive, extrusive, metamorphic, pressure, foliated, non-foliated (BLM #9) (Post BLM #9 in the classroom.)

Graphic organizer (pages 2, 4, and 5 of unit packet), (BLM #10) on which to record observations

Textbook pages describing igneous and metamorphic rocks

Images of volcanoes

Overhead projector with transparency (or any other means of projection) of reading selection about sedimentary rocks

2 pieces of chart paper, one labeled *Igneous Rocks*, one labeled *Metamorphic Rocks*

Connections to Prior Knowledge/Building Background:

- TW tell students that today they are going to be learning about two new kinds of rocks. TW have students play "Oh Yesterday!" (see procedure in Chapter 2) so students recall what they learned yesterday about sedimentary rocks. TW use magic name sticks to call on students to perform.

(continued)

Content Objectives:
SW each make five detailed observations about igneous and metamorphic rocks.

SW identify key characteristics of igneous and metamorphic rocks.

Language Objectives:
SW read for specific information about igneous and metamorphic rocks.

SW orally share five important facts with a partner using sentence frames:

One fact about igneous rocks is . . .

One fact about metamorphic rocks is . . .

- TW tell students that they are going to use their skills to make observations of igneous and metamorphic rocks.
- TW hand out trays of igneous rocks with magnifying glasses. TW call on one student to select a rock and model making an observation to remind the class about writing detailed scientific observations. SW record observations about igneous rocks (on page 2 of the unit packet).
- SW pair-share their observations. Then TW call on students to report out using magic name sticks. TW use the student observations to introduce important facts about igneous rocks. For example, students might notice the crystals in the rocks. TW say "You already know an important fact about igneous rocks. Some igneous rocks have visible crystals. Now we are going to read about igneous rocks to find out what else scientists know about them."
- TW introduce the vocabulary words *lava*, *magma*, *intrusive*, and *extrusive*, using the vocabulary with graphics cards and the volcano images.
- TW read part of the selection aloud and will think-aloud while underlining important information in the text on the transparency. TW model how to put the facts into his/her own words. SW practice underlining by finding one fact in the reading and sharing with the class. TW give specific feedback to students as they underline their first facts. SW finish reading the selection and will underline important facts.
- TW go over language goals and SW state some facts using the sentence frames. They will record the facts in their own words on page 4 of the packet (BLM #10). (See Planning Point 1.)
- After taking notes on igneous rocks, SW use the sentence frame to write one complete sentence on a sticky note. SW pair-share their sentence and correct if necessary. SW place a sticky note on a poster in the front of the classroom under the heading "Igneous Rocks."
- SW repeat this process for metamorphic rocks, including the note-taking activity (on pages 2 and 5 of the packet).
- After taking notes on metamorphic rocks, SW use the sentence frame to write one complete sentence on a sticky note. SW pair-share their sentence and correct it if necessary.

- During each section of the lesson, TW monitor and assess student work by walking around the classroom and reviewing each student's observations and notes.
- TW assess students during think-pair-share and while using the magic name sticks.
- TW assess students by monitoring the room while they are writing their sentences and after they are posted on the charts.

(continued)

SIOP® LESSON PLAN: *Grades 6–8 Unit: Science Rocks, Day 3: Igneous and Metamorphic Rocks*

Developed by Hope Austin-Phillips *(continued)*

Wrap-Up: SW play Pass the Note Card (see procedures in Chapter 3) using the sticky notes. After several rotations, they place the sticky note they end up with on a poster in the front of the classroom under the heading "Metamorphic Rocks." TW review the objectives for today by asking students if the class met all of the goals.

PLANNING POINT 1

As in Day 2's lesson, be prepared to work with a small group of students who aren't ready to complete the task independently.

SIOP® LESSON PLAN: *Grades 6–8 Unit: Science Rocks, Day 4: The Rock Cycle Part I* Developed by Hope Austin-Phillips

Key: SW = Students will **TW** = Teacher will **HOTS** = Higher Order Thinking Skills

Content Standards: Describe the rock cycle.

Distinguish the components and characteristics of the rock cycle for the following types of rocks: igneous, metamorphic, and sedimentary.

Key Vocabulary:
rock cycle, metamorphism, melting, crystallization, weathering, transportation, deposition, compaction, cementation, pressure

HOTS: Applying vocabulary in context

Visuals/Resources/Supplementary Materials:
Igneous, sedimentary and metamorphic rocks

Projected image of the rock cycle diagram (from a textbook or the Internet)

Key vocabulary words with graphics or illustrations on cards

4-Corners Vocabulary Chart (BLM #11)

Animation of rock cycle processes (from textbook or Internet)

Connections to Prior Knowledge/Building Background:

- SW do a 2-minute Quickwrite (see Chapter 2 for Quickwrite procedures) to relate what they have already learned about rocks. SW pair-share their Quickwrite. TW use magic name sticks to call on students to report out. TW review the objectives and tell students that today they will be learning about the rock cycle. TW explain that there are some processes that make the rocks change into other kinds of rocks and they are going to be learning about that. TW hold up a sedimentary rock that the students have seen before and tell the students that at some point the sedimentary rock might become a metamorphic rock or igneous rock. TW hold up one of these other types of rocks for students to see as a reminder of what the different rocks look like.

(continued)

Objectives:	*Meaningful Activities:*	*Review/Assessment:*
Content Objective: SW explore the stages of the rock cycle. **Language Objectives:** SW draw an illustration and write a contextualized sentence to show understanding of the rock cycle vocabulary words. SW orally share definitions of key words with partners.	• TW hand out a 4-Corners Vocabulary Chart to each student (BLM #11). (See Chapter-2 for 4-Corners Vocabulary Chart procedures.) TW introduce each vocabulary word (e.g., *weathering*) by saying the word, explaining what it means, showing a graphic representation of the word, and showing an animation if there is one available. • TW write one word and a definition for the word on the 4-Corners Vocabulary Chart. SW copy the word and definition onto their 4-Corners Vocabulary Chart (or they draw the chart in their notebook). TW model how to draw a graphic representation of the word and will model how to write a contextualized sentence for the first word. SW copy the teacher's example. • TW write a second word and a definition. SW copy the word and definition. SW draw an image of the word and will write a sentence for the word. TW call on volunteers to share their drawings and sentences. TW give specific feedback to correct any errors. • The process will continue until all of the words are completed. (See Think-Aloud 1.) • TW tell students that in the rock cycle, rocks can change into other kinds of rocks. TW project an image of the rock cycle for the students to see. (See Planning Point 1.) TW show students where sedimentary, igneous, and metamorphic rocks are on the diagram. • TW think-aloud to draw and label a rock cycle diagram that students copy in their notebooks. Together they label their diagrams with key vocabulary. • TW remind students of the second language objective and review directions for Milling to Music. (See Milling to Music procedures in Chapter 3.) SW take their 4-Corners Vocabulary Charts with them during the activity. Each time the music stops, TW ask students to share a vocabulary word and definition with their partner.	• TW assess students by monitoring their work on the 4-Corners Vocabulary Charts. • During Milling to Music, TW monitor student responses.

Wrap-Up: Students return to their seats. TW review the objectives for the day and will ask students if they have met them. TW name some key words and SW think-pair-share at least one definition. TW use magic name sticks to call on students to share a response.

(Template developed by Melissa Castillo & Nicole Teyechea. Used with permission.)

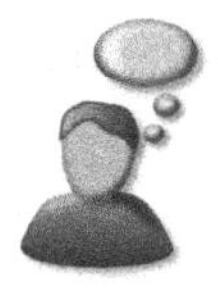

THINK-ALOUD 1

If time is short, I may organize students into groups and assign different words to different students in the groups. They complete the 4-Corners Vocabulary Charts for their words and then share with group members.

PLANNING POINT 1

It is useful to show animations of the rock cycle on the computer. The following site has two useful Web pages: http://www.classzone.com/books/earth_science/terc/content/investigations/es0602/es0602page02.cfm and http://www.classzone.com/books/earth_science/terc/content/investigations/es0602/es0602page03.cfm.

SIOP® LESSON PLAN: *Grades 6–8 Unit: Science Rocks, Day 5: The Rock Cycle, Part II* Developed by Hope Austin-Phillips

Key: SW = Students will **TW** = Teacher will **HOTS** = Higher Order Thinking Skills

Content Standards: Explain the rock cycle.

Distinguish the components and characteristics of the rock cycle for the following types of rocks: igneous, metamorphic, and sedimentary.

Key Vocabulary:	**Visuals/Resources/Supplementary Materials:**
rock cycle, metamorphism, melting, crystallization, weathering, transportation, deposition, compaction, cementation, pressure **HOTS:** Applying vocabulary; interpreting diagrams	Igneous, sedimentary and metamorphic rocks Projected image of the rock cycle diagram (on a transparency or via computer) Image of the rock cycle without labels for student to label (one copy per student) Vocabulary words with graphics (BLM #9). This can be posted in the classroom. Rubric for rock cycle letter (BLM #12)

Connections to Prior Knowledge/Building Background:

- TW instruct students to take out their 4-Corners Vocabulary Charts from the prior lesson. TW instruct students to review the vocabulary words with their groups. One student will read a definition while all others listen and try to guess the word. Another student will read a definition while others listen and guess. The students continue sharing around the table until all of the words have been reviewed. TW review the words with the class.
- TW tell students that today they are going to use these vocabulary words in a writing activity about the rock cycle. TW go over the objectives.

(continued)

Objectives:	*Meaningful Activities:*	*Review/Assessment:*
Content Objective: SW apply knowledge to recreate the rock cycle. **Language Objectives:** SW use new vocabulary words to write a letter to a family member pretending to be a rock stuck in the rock cycle. SW use sequence words to tell a story about the rock cycle: First . . . Next . . . Then . . . After that . . . Finally . . .	• TW ask the students questions about the rock cycle such as, "How could a metamorphic rock turn into an igneous rock?" SW think-pair-share responses. • TW distribute a rock cycle diagram without labels and have students work in groups to complete it. • TW tell students that now they are going to apply their new knowledge of the rock cycle. TW tell students that they are going to pretend to be a rock in the rock cycle. They will write a letter to their family telling about their journey through the rock cycle. SW include all of the vocabulary words in their letter. • While introducing the assignment, TW discuss how a friendly letter is organized, including the sections such as date, greeting, body, and closing. TW review the sequence words to use. • SW write their letters to their family and include vocabulary and sequence words.	• TW monitor and assess students during the think-pair-share activity. • TW monitor students as they write the letter to their family. • TW assess student letters for accurate explanations of the rock cycle and use of vocabulary and sequence words. (See Planning Point 1.) TW use the rubric (BLM 12).

Wrap-Up: SW circle one paragraph from their letter they would like to share. First, they pair-share their paragraph. Then TW use magic name sticks to call on students to share their selection of their letter. TW review the objectives for the day and ask students if they have met all of the goals.

(Template developed by Melissa Castillo & Nicole Teyechea. Used with permission.)

PLANNING POINT 1

You may want to post a Signal Words poster of sequence terms on your wall (see procedures in Chapter 2). The five sequence words in the language objective are quite basic. For intermediate and advanced students, you may want to encourage them to use other sequence terms.

Concluding Thoughts

Middle school students like to be active! This unit gives them several opportunities for discovery learning and also teaches them the distinction between an observation and a fact. The unit packet is a handy resource for the students to organize the information they learn about these rock types, and the language scaffolds ensure they will report findings in a scientific way. The final writing task requires them to synthesize their knowledge in a creative way, imagining they are a rock and describing a journey through a letter home.

Extension activities can be planned for more advanced students and/or the curious. One idea is to categorize rocks by hardness. Students can bring in their own rocks as well as those provided by the teacher to create a real index of hardness. Students are always excited to learn that a diamond is the hardest rock of all, but it comes from coal, a relatively soft rock. Your students will probably never look at a rock in the same desultory way again!

We hope this unit gave you new ideas for incorporating meaningful activities and attention to language development in your science lessons. We encourage you to read Chapters 5, 6, and 8 for additional unit lessons. Even if the grades you currently teach focus on grades 6–8, you will find interesting SIOP® science lessons and effective integration of the new techniques in these other sample units.

chapter 8

SIOP® Science Units and Lessons for Grades 9–12

Hope Austin-Phillips and Deborah Short

Introduction

This chapter presents a five-day science unit on the cell, including lesson plans and handouts. While it is designed for high schoolers, we recognize that some states cover this material in their middle school standards as well. You should be able to adapt these lessons for those younger learners. Hope Austin-Phillips, one of our SIOP® science educators, begins this chapter by describing in detail how she designed this unit, with comments on the decisions she made in order to write lessons aligned to the SIOP® Model. She offers her insights for planning one particular lesson within this unit, Day 1. (Note: If you have

not read Chapter 4 about SIOP® science unit and lesson design for English learners, please read it for an overview of the planning process and a discussion of the lesson plan format used in this and other chapters.) As you read this unit, you will see how several of the techniques from Chapters 2 and 3 are incorporated into the lessons, as well as how the activities in the unit lesson plans map onto the components and features of the SIOP® Model (found in Appendix A).

You will notice that there are Think-Alouds and Planning Points in many of the lessons. The Think-Alouds, identified with a thought bubble icon, are questions we asked ourselves and/or decisions we made as we undertook the SIOP® unit planning process. The purpose of the Planning Points, identified with a flash drive icon, is to highlight or identify resources that support the implementation of the lesson.

Cells

Unit Overview, Grades 9–12

Vocabulary

When introducing the cell to English learners, it is very important to emphasize key vocabulary. Many of the words are new terms that the students have no connections with. I make sure that the words are posted, discussed, and repeated many times. If there is a word that is particularly hard to pronounce, I tell students that in order to leave the classroom, they need to say the word on their way out the door. That way, all of the students say the word at least once and the students who are unsure of the word can hear many other students say the word before they have to say it. This makes them more comfortable with the vocabulary.

Hands-On Experiences

I introduce this unit with a hands-on experience. Giving students the chance to see real cells under a microscope gives them an immediate connection to the material. Because the students can easily see the cell wall, cell membrane, cytoplasm, and nucleus, they make much stronger connections with these vocabulary words. I teach the vocabulary in more detail after students have examined the cells under the microscope. Some of my beginning students with interrupted educational backgrounds haven't used microscopes, slides, and cover slips before, so I help them use these tools and assign more knowledgeable peers to partner with them.

Interaction

In this unit, I use a variety of interaction techniques to keep the students engaged. For one, I build in many opportunities for students to Think-Pair-Share, which is important because that technique allows more wait time and lets the students rehearse before they are asked to share an answer with the whole class. I never expect a student to answer a question in front of the class *unless* I have given him or her a chance to discuss the topic with other students and ask questions. After students have shared with another student, I expect that they can respond to the prompt in front of the class.

I use another interaction technique in this unit when the students are learning about the cell organelles. In this activity, they take turns reading the paragraphs aloud to their partners. After each paragraph is read aloud, the students have to summarize the information to put it on their graphic organizer. By allowing the students to work in groups for this activity, I give them a chance to discuss and share the important points before they write their summary. This process ensures that students will have more clear, concise answers on their paper. This technique is so effective and meaningful to students that I incorporate it in two days of this unit.

I also use Milling to Music, a great interactive procedure for kinesthetic learners and students who do not like to stay in their seats! In this activity, they have a chance to speak with several partners in a relatively short time frame while they are practicing the necessary language for the lesson.

Reading Material

The reading material for this unit about cell organelles can be found in any high school biology textbook. For lower-level English learners, I suggest using simple sentence strips instead of paragraphs. I have included the sentence strips I use with my lower-level ELs. When using the strips, the students are only looking for information about the function of the organelles, not the structure. These students will get their information about structure from the cell diagrams.

The Cell Project

The cell project has been one of the most enjoyable projects I have ever done with my students. In my own high school biology class, I did a cell project and still remember it today. I learned so much from it. Every year when I introduce the project, I start second-guessing myself and wonder if so much class time should be spent on a big project like this, but my students more than astound me! This past year, some students created amazing raps, others hilarious skits, and still others gorgeous scrapbooks and some very interesting posters. In my opinion, every minute of class time spent on the project is well worth it.

I usually introduce the cell project one day and then give the students three or four more days in class to work on it. If time is limited, the project can be done as homework. I highly recommend spending at least a couple of class days on it because that gives you time to conference with each student group. I find that the conferencing process gives me a chance to help the students if they are stuck and to informally check their comprehension of plant and animal cells. It also helps me reinforce the fact that their project is being assessed with the scoring rubric. This rubric helps to keep the students focused. When I meet with the groups, I first make sure they have included the structure and function of each organelle. I also verify that they have used all of the vocabulary words in context. At the end, I have each group orally present their project to the class so all students can learn about the other projects.

Further Investigation into the Cell

The lessons shared in this unit are meant to be introductory lessons to the cell. As part of a rigorous science curriculum, the cell processes will be further explored with other laboratory experiments and reading exercises. This unit represents a critical base of knowledge for later topics such as cellular transport, energy production, reproduction, and more.

Now, let me explain to you some of the planning process I undertook for this unit. I'll describe my deliberations as I developed the first lesson.

Day 1 Lesson Design: Cells

Beginning a unit about cells is always an exciting time for me. Students can be very interested in cells, and because cells are so easy to see with microscopes, students make immediate connections to the information. In planning the first day of the cell unit, my main goal is for students to have some hands-on experiences with cells. My objectives are:

- Students will prepare a wet mount slide of cheek cells and onion skin cells and observe under a microscope.
- Students will write observations and draw diagrams of the cells seen under the microscope.

I chose these objectives because I want students to gain exposure to as many lab techniques as possible during the course of the unit. Creating a wet mount slide is a lab technique that is essential to life science classes from middle school through college. Further, creating a wet mount slide of cheek cells and onion skin cells is very simple and will allow students to see real cells.

I also want students to practice making detailed observations of cells. Students should write observations as well as draw detailed diagrams. Students need to practice looking closely and carefully so they can notice the fine details. If students look closely at cheek cells and onion skin cells, they will be able to see cell walls (in the onion skin cells), cell membranes, cytoplasm, and nuclei. Once they have seen these structures in the cells, they are able to connect to the vocabulary words when I introduce them later in the lesson.

My language objective for the day is:

- Students will orally tell a partner the steps to creating a wet mount of cheek cells and onion skin cells using sequence words: First, _____; Second, _____; Third, _____.

When the students are conducting a lab, I will often have a language objective similar to this because it helps the students process the directions for the lab. It can be overwhelming to look at lab directions, so I like the students to break them down into steps. When the students tell a partner the steps using sequence words, it helps them understand what they need to carry out before the experiment starts. This objective is also helpful for me to assess whether they understand the directions. I call on several students to state the steps of the experiment after they talk with a partner. If many students are struggling with this language objective, I know that I need to review the directions again and perhaps model each step before the students begin.

After figuring out my objectives, I decide what activities will align with my objectives. Because this is an introductory lesson on cells, I want to tap into the students' background knowledge before starting the lesson. I have found that a lot of students have very little background knowledge about cells, and often, even if they have some background knowledge, it takes them a while to remember it! So, I like to start this lesson with a Quickwrite because it gives the students a few minutes to write down everything they know about cells. Before we do the Quickwrite, however, I want students to have some exposure to cells to help them "remember" what they already know. I may have the students do a short reading assignment the night before the lesson, or I may show students a short video clip about cells or some images of cells. After one of these reminders, I will have them complete the Quickwrite. In this way, they can write what they know about cells or questions they have about cells. The prompt thus allows all students to participate. Students who have some background knowledge can write what they know about cells, and students with very little background knowledge can write their questions. I read the Quickwrites as soon as possible so I can have a better understanding of the breadth of my students' knowledge (or lack of knowledge) about cells.

After tapping into their background knowledge, I focus on the lab activity. Unfortunately, I find that many students come into my class with little or no experience using microscopes, so it is very important to make sure that students understand the techniques used in the lesson. To do this, I post the Procedures with Graphics poster as a resource. First, I read the Procedures with Graphics poster and model the steps of the experiment. Then, I have the students read the directions and restate the directions to their partner using sequence words. This allows me to check for understanding. When I am sure that students understand the directions, I let them complete the lab activity. Students often need a little help determining what they are looking for with the microscope, but then get very excited once they realize they have found cells! I am fortunate enough to have a microscope camera in my room and I project images of the cells for students to see a few minutes after the lab has started. This step is helpful because then the students know what to look for and can begin their observations and diagrams.

After they have completed the lab activity, I introduce the vocabulary to them. I explain *cell wall, cell membrane, cytoplasm*, and *nucleus*. Students are all usually quite interested in learning the words because they have already seen examples of these cell organelles in real life! Finally, for a review of the terms and the observations they made, the students participate in a Simultaneous Round Table activity. This allows them to quickly remember and review what they have learned during the lesson. I will then call on students to share their new knowledge and I make sure to highlight the new vocabulary words.

9–12 Unit

The unit in this chapter addresses a major biology topic, cells. As you read the lessons, consider how the activities that Hope Austin-Phillips designed enable students to reach deeper levels of understanding about these microscopic cells and their organelles. Some students worksheets are listed as BLMs (Blackline Masters) and are located in Appendix C.

SIOP® LESSON PLAN: *Grades 9–12 Unit: Cells, Day 1: Cheek Cell and Onion Skin Cell Lab* Developed by Hope Austin-Phillips

Key: **SW** = Students will **TW** = Teacher will **HOTS** = Higher Order Thinking Skills

Content Standard(s):
Compare and contrast plant and animal cell structures and functions.

Key Vocabulary:
glass slide, coverslip, cell membrane, cell wall, nucleus, cytoplasm, plant cell, animal cell

Visuals/Resources/Supplementary Materials:
Video clip or images about cells (See Planning Point 1).
Procedures with Graphics poster (See Planning Point 2).

(continued)

SIOP® LESSON PLAN, *Grades 9–12 Unit: Cells, Day 1: Cheek Cell and Onion Skin Cell Lab* Developed by Hope Austin-Phillips *(continued)*

	Microscopes
	Supplies for lab including: toothpicks, methylene blue dye, glass slides, coverslips, water dropper bottle, 1–2 onions
HOTS: Preparing a wet mount slide; identifying cell structures.	Cell Lab report form for recording observations and diagrams (BLM #13)

Connections to Prior Knowledge/Building Background:

- Often students have very little prior knowledge about cells, so it is helpful to show a short video clip about cells. If it is not possible to find a video clip, show students some images of different types of cells and reinforce the idea that all living things are made of cells. After the video clip (or other images), SW complete a 3-minute **Quickwrite** about cells (see procedures for Quickwrite in Chapter 2). The prompt for the Quickwrite is: "Write for three minutes. Discuss what you know about cells and list any questions you have."
- TW ask for student volunteers to share what they know about cells and the questions they may have. TW tell students "Today we are going to look at some real cells!" TW review content and language objectives for the day.

Objectives:	*Meaningful Activities:*	*Review/Assessment:*
Content Objectives: SW prepare a wet mount slide of cheek cells and onion skin cells. SW observe cells and draw diagrams of those seen in the microscope. **Language Objective:** SW orally tell a partner the steps for creating a wet mount slide of cheek cells and onion skin cells using sequence words: First . . . Second . . . Third . . .	• TW introduce directions to create a wet mount slide by referring to posted Procedures with Graphics (see explanation in Chapter 2). SW read the procedures for creating a wet mount slide of cheek cells aloud with a partner. TW use magic name sticks (craft sticks with student names) to call on students randomly to share procedures one step at a time. TW model each step and refer to the posted procedures with graphics.	
	• SW create a wet mount slide of cheek cells. (See Think-Aloud 1) SW observe cells under high power with the microscopes. SW draw a detailed diagram and write observations of the cells on the Cell Lab report form (BLM #13).	• TW monitor students as they create their wet mount slides.
	• SW read the procedures for creating a wet mount slide of the onion skin cells aloud with a partner. TW call on students to share each step for making the wet mount slide of the onion skin cells and will remind them to use sequence words. TW model each step and refer to the posted procedures with graphics.	• TW use magic name sticks to call on students to share the steps with the class.
	• SW create a wet mount slide of the onion skin cells. SW observe the cells, draw a diagram, and write three detailed observations on the Cell Lab report form.	
	• SW do a think-pair-share to discuss their observations. TW record observations on the board. TW introduce the vocabulary words: *plant cell, animal*	• TW give positive and informative feedback to students as they label their diagrams.

(continued)

cell, cytoplasm, cell membrane, cell wall, and *nucleus*. TW think-aloud while labeling one part of the cell. SW label the rest of the structures on their drawings.

Wrap-Up: SW participate in Simultaneous Round Table (see procedures in Chapter 3) to review their observations and vocabulary for the day. TW call on student volunteers to share their comments. TW highlight any vocabulary words that are brought up and will remind students of any vocabulary that was not brought up. TW review the goals for the day.

(Template developed by Melissa Castillo & Nicole Teyechea. Used with permission.)

PLANNING POINT 1

Many commercial textbooks have CDs or websites with relevant video clips. Also, your school may have a subscription to Brainpop or Discovery Education, both of which have good video clips on cells.

PLANNING POINT 2

I prepare Procedures with Graphics posters for all the science labs. I list the steps and draw pictures to remind students what to do. If students forget what to do next, they can refer to the poster and not have to question me. I can also re-use these posters each year, especially if I laminate them.

THINK ALOUD 1

Because some of my English learners have interrupted educational backgrounds, not all may be familiar with the procedures for using a microscope or preparing a wet mount slide. In these cases, I partner these ELs with more experienced peers.

SIOP® LESSON PLAN: *Grades 9–12 Unit: Cells, Day 2: Animal Cells* Developed by Hope Austin-Phillips

Key: **SW** = Students will **TW** = Teacher will **HOTS** = Higher Order Thinking Skills

Content Standard(s):
Compare and contrast plant and animal cell structures and functions

Key Vocabulary:
animal cell, structure, function, cell membrane, cytoplasm, nucleus, nuclear membrane, nucleolus, mitochondria, golgi apparatus, rough endoplasmic reticulum, smooth endoplasmic reticulum, ribosomes, lysosomes

HOTS: Summarizing structure and function of cell organelles.

Visuals/Resources/Supplementary Materials:
Vocabulary words with graphics (See Planning Point #1.)

Images of cells and cell organelles

Organelles definition sentence strips (BLM #14) (See Think Aloud 1.)

Structure/function chart to record information (BLM #15)

Diagram of the animal cell with organelles labeled

(continued)

SIOP® LESSON PLAN: *Grades 9–12 Unit: Cells, Day 2: Animal Cells* Developed by Hope Austin-Phillips *(continued)*

Connections to Prior Knowledge/Building Background:

- TW tell students that today they will be learning more information about cells. TW ask students to think-pair-share things they noticed during the lesson yesterday. TW go over the objectives for the day and tell students that today they will be learning more about animal cells.

Objectives:	*Meaningful Activities:*	*Review/Assessment:*
Content Objective: SW explain the structure and function of at least 8 animal cell organelles.	• TW introduce the words *structure* and *function*.	
Language Objective: SW listen to information and record summary notes.	• TW tell students that they will be learning about the different structures inside an animal cell. SW think-pair-share the structures they could see in their cheek cells. TW introduce the new vocabulary words using the graphics.	• TW assess students while they are working on their graphic organizers. SW review this objective while completing the language objective activity.
SW use the following sentence frames to describe the structure and function of each animal cell organelle: The structure of the ______ is __________. The function of the ______ is __________.	• TW form student groups and hand out an envelope to each one. The envelopes contain a short sentence or paragraph strips about the structure and function of different cell organelles (see BLM #14). TW review directions for the activity: SW open the envelope. SW take turns reading aloud the sentence strips to their group. After the sentence has been read, SW summarize the important information on their graphic organizer (BLM #15).	
	• Before students start, TW will read aloud the paragraph about the nucleus. TW think-aloud while summarizing the important information to model how to record on the graphic organizer.	
	• One student will read aloud the strip for cell membrane to the group. That student will summarize important information and record on the graphic organizer, with help from peers as needed. The others record the summary too. SW share with the class and TW give specific feedback to students as they share their summaries.	
	• Student groups will complete the rest of the graphic organizer using this process. (See Think-Aloud 2.)	
	• After completing their graphic organizers, SW choose two organelles and will complete the sentence frames from the language objective. SW write the two sentences on a small piece of scratch paper. SW read their sentences to a partner.	• TW circulate as students read their sentences to a partner.

(continued)

Wrap-Up: Toss-an-answer: TW call on students to share one of their sentences with the class. After the students read their response, they crumple up their paper and toss it into a bucket the teacher has. TW call on 5–10 students and will then take volunteers. TW review the goals for the day and ask students if they have met the objectives. TW collect the envelopes with the sentence strips inside.

PLANNING POINT 1

I prepare charts or word cards of key vocabulary with graphics for this unit. I post the chart or display the cards throughout the unit.

THINK-ALOUD 1

I cut apart the paragraphs on BLM #14 to make sentence strips. I place them in an envelope and make enough sets for the small groups. I like to use these strips because they are more comprehensible than the textbook for my lower proficiency level students. It is a good way to introduce the important organelles.

THINK-ALOUD 2

The first time students do the read-listen-record activity, it often takes more than one class period so the groups may need to finish recording information about some of the animal cell organelles in the next lesson.

SIOP® LESSON PLAN: *Grades 9–12 Unit: Cells, Day 3: Plant Cells* Developed by Hope Austin-Phillips

Key: SW = Students will **TW** = Teacher will **HOTS** = Higher Order Thinking Skills

Content Standard(s):
Compare and contrast plant and animal cell structures

Key Vocabulary:	**Visuals/Resources/Supplementary Materials:**
plant cell, cell wall, cell membrane, cytoplasm, nucleus, nuclear membrane, nucleolus, mitochondria, golgi apparatus, rough endoplasmic reticulum, smooth endoplasmic reticulum, ribosomes, lysosomes, chloroplasts	Vocabulary words with graphics Images of cells and cell organelles Organelles definition sentence strips (BLM #14) Structure/function chart to record information (BLM #15)
HOTS: Summarizing information.	Diagram of a plant cell with the organelles labeled

Connections to Prior Knowledge/Building Background:

- SW review the structure and function of the organelles in an animal cell. TW review the procedures for Milling to Music. (See procedures for Milling to Music in Chapter 3.) During Milling to Music, TW name an organelle and SW tell their partner the function of the organelle or TW name a function and SW name the organelle.

(continued)

SIOP® LESSON PLAN: *Grades 9–12 Unit: Cells, Day 3: Plant Cells* Developed by Hope Austin-Phillips *(continued)*

Objectives:	*Meaningful Activities:*	*Review/Assessment:*
Content Objective: SW describe the structure and function of at least eight plant cell organelles.	• TW review the words *structure* and *function*. TW tell students that they will be learning about the different structures inside a plant cell. SW think-pair-share the structures they could see in the onion skin cells. TW introduce the new vocabulary words.	
Language Objectives: SW use the following sentence frames to describe structure and function of each plant cell organelle: The structure of the ______ is ______. The function of the ______ is ______.	• TW hand out envelopes with sentence strips (BLM #14) containing information about the structure and function of different plant cell organelles. TW review directions for the activity: "This will be similar to yesterday's lesson, but we will also compare plant and animal cells."	
SW use comparative language to discuss similarities and differences among plant and animal cells.	• SW open the envelope. Students will take turns reading aloud the sentence strips to their group. After the sentence has been read, SW record the information on their graphic organizer.	• TW assess students while they are working on their graphic organizers. SW review this objective while completing the language objective activity.
	• Before students start, TW will read aloud the paragraph about the nucleus. TW think-aloud while summarizing the important information to write on the graphic organizer. TW remind students that this is the same activity they did yesterday, but they will be learning some new organelles and will be reviewing some organelles.	
	• SW complete the rest of the graphic organizer (BLM #15)	
	• After completing their graphic organizers, SW choose two organelles and will complete the sentence frames from the language objective. SW write the two sentences on a small piece of scratch paper. SW read their sentence to a partner.	• TW monitor while the students read their sentences.
	• TW review comparative language frames, as needed, with the students and have students think-pair-write-share two comparisons between plant and animal cells. SW add one comparison sentence to their scratch paper.	

(continued)

Wrap-Up: Pass the Note Card: Using their scratch paper as the note card, students stand in a circle and pass their papers while the music plays (see procedures in Chapter 3). When the music stops, TW call on some students (e.g., all those wearing green) to read the comparison sentence on the paper they are holding. TW review the goals for the day and ask students if they have met the objectives.

(Template developed by Melissa Castillo & Nicole Teyechea. Used with permission.)

SIOP® LESSON PLAN: *Grades 9–12 Unit: Cells, Day 4: Comparing and Contrasting Plant and Animal Cells*

Developed by Hope Austin-Phillips

Key: SW = Students will **TW** = Teacher will **HOTS** = Higher Order Thinking Skills

Content Standard(s):
Compare and contrast plant and animal cell structures.

Key Vocabulary:
animal cell, plant cell, cell wall, cell membrane, cytoplasm, nucleus, nuclear membrane, nucleolus, mitochondria, golgi apparatus, rough endoplasmic reticulum, smooth endoplasmic reticulum, ribosomes, lysosomes, cholorplasts

HOTS: Comparing; contrasting

Visuals/Resources/Supplementary Materials:
Vocabulary words with graphics

Snowball! form (BLM #16)

Venn diagram (See Planning Point 1)

Comparative Signal Words poster (BLM #16)

Chart paper with two columns, headed "Structure" and "Function" posted on board or wall

Connections to Prior Knowledge/Building Background:

- Snowball! SW write facts they have learned about plant and animal cells on their Snowball! form (BLM #16) (see directions for Snowball! in Chapter 3). After the snowball activity, TW use magic name sticks to randomly call on students to share what they have learned. TW review the objectives for the day.

Objectives:	*Meaningful Activities:*	*Review/Assessment:*
Content Objective: SW visually represent plant and animal cell structures. SW categorize information about the structures and functions of the two types of cells using a Venn diagram. **Language Objective:** SW use comparative language terms and phrases to state the similarities and differences between animal and plant cell structure and function to a partner.	• TW refer back to the lab activity in which the students observed both human cheek cells and onion skin cells. TW explain that today, they are going to be looking deeper into the similarities and differences of the animal and plant cells. • SW draw diagrams of an animal cell and a plant cell in their notebooks and label the cell organelles. Then they draw and complete a Venn Diagram to compare and contrast plant and animal cells. • TW review and model comparative language with the students using the Compare/Contrast Signal Words poster (BLM # 17) • SW think-pair-share the similarities and differences between the cheek cells and onion skin cells from the	• TW check on the students' labeled drawings of animal and plant cells and their Venn Diagrams.

(continued)

SIOP® LESSON PLAN: *Grades 9–12 Unit: Cells, Day 4: Comparing and Contrasting Plant and Animal Cells*

Developed by Hope Austin-Phillips

Objectives:	*Meaningful Activities:*	*Review/Assessment:*
	lab and try to use the signal words and sentence frames, such as:	
	One similarity between animal and plant cell structure/function is______. One difference between animal and plant cell structure/function is________.	• TW use magic name sticks and have students share some sentences.
	• After oral practice, SW write three comparative sentences on a sticky note. SW read their sentences to a partner.	• SW place their sticky notes on the poster under the category of "Structure" and "Function" as appropriate.

Wrap-Up: TW review the goals for the day. SW raise one finger if they do not think that they mastered the goals for the day. SW raise two fingers if they partially understand the goals for the day but need a little more practice. SW raise three fingers if they feel they fully mastered the content and the comparative language structures.

(Template developed by Melissa Castillo & Nicole Teyechea. Used with permission.)

PLANNING POINT 1

The teacher could provide an outline of a Venn diagram to every student. However, this graphic organizer is relatively easy for students to draw on their own, as is done in this lesson.

SIOP® LESSON PLAN: *Grades 9–12 Unit: Cells, Days 5–7: The Cell Project*

Developed by Hope Austin-Phillips

Key: SW = Students will **TW** = Teacher will **HOTS** = Higher Order Thinking Skills

Content Standard(s):
Compare and contrast plant and animal cell structures.

Key Vocabulary:
eukaryote, eukaryotic cell,animal cell, plant cell, cell wall, cell membrane, cytoplasm, nucleus, nuclear membrane, nucleolus, mitochondria, golgi apparatus, rough endoplasmic reticulum, smooth endoplasmic reticulum, ribosomes, lysosomes, chloroplasts

HOTS: Comparing; contrasting; summarizing

Visuals/Resources/Supplementary Materials:
All visuals and resources from previous four lessons

Any available books about cells

Internet access to websites about cells (if possible)

Cells Project rubric (BLM #18)

(continued)

Connections to Prior Knowledge/Building Background:
TW tell students "you are now going to have a chance to show everything you have learned about plant and animal cells."

TW ask students to think-pair-share something that they have learned that surprised them about plant and animal cells. (See Think-Aloud 1.)

TW call on students to share answers. TW introduce the objectives for the day.

Objectives:	*Meaningful Activities:*	*Review/Assessment:*
Language Objectives: SW create a project that demonstrates understanding of plant and animal cell organelles. **Language Objectives:** SW use the content vocabulary in context in their project. SW use proper spelling and grammar in their project.	• TW tell students that over the next several days, they will be creating a "Cell Project" that demonstrates their understanding of plant and animal cells. (See Think-Aloud 2.) • TW go over the rubric (BLM #18) with the students. TW provide a quick review of the vocabulary words while discussing the rubric. • TW tell students that they need to meet the requirements on the rubric, but they can pick any format they want to demonstrate their understanding of cells. TW show students a list of possible projects that include: write and perform a rap or a song; create and present a PowerPoint presentation; create and present a scrapbook; write and perform a skit; create and present a poster; create and present a comic strip; create a picture book; or create a model. • TW encourage students to use resources at their disposal in the room and on the Internet.	• TW conference with each team every day that the students work on the project to assess their progress on the rubric. • Once all projects are done, each student or group presents its poster. Each project will be assessed using the rubric (BLM #18).

Wrap-Up: During the work days on the posters, TW ask students or groups to share their plans for their project with the class so all class members have an idea of what the other students are doing. (See Think Aloud 3.)

During the days that students present their posters, SW respond to one or two of these questions on an index card (e.g., How did things go? What would they do differently? What did you learn about teamwork? What did you learn about cells?). TW collect these as Tickets Out.

(Template developed by Melissa Castillo & Nicole Teyechea. Used with permission.)

THINK-ALOUD 1

As teachers, we regularly call on students to recall facts. Here, instead, I add a twist. Using higher-level thinking, students have to determine what they learned that surprised them about cells.

THINK-ALOUD 2

For this kind of project, I usually give the students a choice. They can (a) work alone, or (b) tell me three people they would like to work with. Then I put them in groups and do my best to get them with at least ONE of the classmates they wanted to work with.

THINK-ALOUD 3

Because these projects take several days to complete, I select different groups that volunteer to present their project designs at the end of each lesson. I ask volunteers to briefly tell the class what their project is going to be in hopes that it might inspire other students to be creative. For example, one of my groups this year did the "Diary of Golgi Apparatus." I had them quickly explain to the class what they were doing to hopefully give kids some new ideas. For students with lower English proficiency levels, I offer them the chance to share their ideas earlier when the environment feels more informal.

Concluding Thoughts

Understanding cells is a critical stepping stone for much of the biology curriculum. Indeed, knowledge of cells, cell organelles, and their functions is important for other science courses and for adults who want to be scientifically literate in our society. Many modern scientific findings—from cancer treatments to the human genome project—are based in the concepts presented in this unit. Therefore, it is essential that students have access to this information in ways that make it comprehensible to them. Further, our English learners need the language scaffolds described in this unit to articulate their hypotheses, observations, conclusions, and applications. Together, the activities for content and language practice advance their understanding and scientific language skills.

We hope this unit gave you new ideas for incorporating meaningful activities and attention to language development in your science lessons. We encourage you to read Chapters 5, 6, and 7 for additional unit lessons. Even if the grades you currently teach focus on grades 9–12, you will find interesting SIOP® science lessons and effective integration of the new techniques in these other sample units.

chapter 9

Pulling It All Together

As we planned this final chapter, we decided to share some of the things we have learned in the process of writing this book, related to our collaborations with our content contributors and our understandings of lesson and unit planning using the SIOP® Model. We also asked our contributors to share what they have learned, and we have included their thoughts and insights in the second half of this chapter. Together, we have learned to be better SIOP® teachers and professional developers.

What We Have Learned

One important finding for all of us is the confirmation that becoming an effective SIOP® teacher is a process that takes time, reflection, practice, and commitment. Unlike many of the educational initiatives that we have all been involved in during our careers, the

SIOP® Model is not about tweaking our teaching a little here and adding a little something there, while expecting immediate results in our students' academic achievement. Instead, the SIOP® Model is about purposeful planning, consistent attention to teaching the academic language and content of your discipline, and maintaining the belief that all students, including English learners, can reach high academic standards while developing their English proficiency. We know from our SIOP® research studies that if teachers are high implementers of the model, their students' academic performance increases significantly (Center for Applied Linguistics, 2007; Echevarria, Richards, Canges & Francis, 2009; Short, Fidelman & Louguit, 2009).

We also have become even more aware that good teaching is about attention to detail. For example, as we were reading and editing the lesson and unit plans created by our contributors, from time to time we had to call or email and ask questions about the purpose of a particular handout, the steps to a process, or the application of an activity. This made us realize how important it is to use precise language with English learners, both in our speech and in the materials we prepare for them. Consistent labeling for classroom routines, procedures, and activities reduces ambiguity and confusion, and serves as additional scaffolding for ELs. How great it would be if all teachers across the school used the same lexicon for their activities! With the SIOP® Model and collaboration, it is possible to do just that. Then when students walk into any class and are told to prepare Cornell notes, for instance, they will know immediately what to do. Clear task explanations allow students to get down to work quickly and facilitate classroom management.

We also had to seriously consider the role of teachers as content experts. You know where you are going with a lesson, and what needs to be taught, learned, and assessed. Students, including English learners, don't have this insider information, and sometimes they are academically lost because they do not know what is expected of them, nor what they are to do. Obviously, this is a primary function of the content and language objectives, to point the way and assist students in knowing what to expect. But when we make assumptions about what students know and can do, we may be basing those assumptions on what *we* know and what *we* can do. Being precise in your use of terms, and carefully explaining and modeling processes and procedures related to the content you are teaching will assist your English learners in becoming more successful in learning your content.

We also sharpened our skills in designing SIOP® lessons and units. Our contributors, for the most part, created and wrote the lessons based on their teaching experiences. Our role was to clarify, verify, elaborate, and expand—adding more ideas, differentiations, and the like. In doing this, we became even more aware of how challenging it is to write detailed SIOP® lesson and unit plans. Teachers new to the SIOP® Model often balk at the time and work it takes to write lessons, yet as we have mentioned previously, with practice, the amount of time and effort is diminished. The end goal needs to be kept in sight: the academic and language proficiency benefits for the English learners who will be productive members of our society in the future. Our contributors to the ELA book explained, "As we gained more experience and more practice using the SIOP® Model, it just became an internalized part of our natural teaching style. We're not denying that it was hard work and we experienced successes as well as failures along the way, but we realized that, like with anything in life, once you begin to reap the benefits of a challenging task, the challenges don't seem so immense" (Vogt, Echevarria, & Short, 2010).

Finally, we have learned from our contributors that experienced, knowledgeable, successful teachers who are well-versed in the SIOP® Model continue to grow and learn through the process of carefully planning and teaching effective and appropriate lessons for English learners. As one of the contributors to the math book explained, "Coming up with higher-order thinking questions (consistently) was difficult for me. Therefore, I know that I need additional professional development in this area. Self-awareness is the key to change. This project helped me identify some of my own weaknesses" (Echevarria, Vogt & Short, 2010).

In the section that follows, you will hear the voices of our SIOP® science specialists. They elaborate on what they learned during the process of working together to help write this book.

What Our SIOP® Science Contributors Have Learned

Hope Austin-Phillips

During this process, I have come to greatly appreciate collaboration. While I enjoy working on my own, there are times when I need a second opinion about something. During the process of working on this book, there were moments when I thought I had reached a dead-end and would have to completely start over. After talking with my partner, I would come away with fresh ideas and ways to improve what I was already working on.

Working on this book has also helped me realize how much work it is to write a book! I am very thankful that I am a contributor and not a main author. I am extremely grateful to have had the opportunity to participate in this process as I have learned a tremendous amount. And in case I ever decide to write my own book, I will have some idea as to what it entails!

Finally, while I feel that I do have a great background for working with English learners using the SIOP® Model, I have also discovered that I have a lot to learn. I wish I could do student teaching all over again (well . . . not really), and visit many classrooms that are using the SIOP® Model to get a wider view of what other teachers are doing to best serve their students. I feel like I get stuck in a rut doing the same activities over and over again. What I want now is to communicate with other teachers to find out what they are doing that I might like to incorporate into my classroom.

Amy Ditton

As teachers, we tend to follow patterns as we plan and deliver instruction. Even if our patterns include high-quality instruction for our English learners, we have to take a step back and reflect on whether we are including enough variety in order to move students through the stages of language proficiency. For example, I have a tendency to focus on speaking and writing in my language objectives. I need to be sure to include listening and reading goals as well, on a regular basis.

Sometimes we think we are being very clear as teachers, when in actuality our message may not be straightforward to others. As I read through my lesson plans, I knew exactly what I was trying to say. However, when I received feedback, my colleagues had many questions that helped me realize that my message was not as transparent to others. It made me wonder: If my plans weren't clear to my colleagues, how clear would my lesson be to students?

This question leads me to a third point. Collaboration was essential for me. It was difficult to sit alone and plan and put my thoughts into writing. The questions that my colleagues asked and answered led to more clearly planned lessons with a variety of objectives and techniques included.

Final Thoughts

As you read the thoughts and insights of our contributors, you may have noticed that they mentioned the importance of collaboration. We couldn't agree more. For over fifteen years, we have collaborated about the SIOP® Model, with each other and with educators throughout the country and now throughout the world. We are convinced that these collaborations have resulted in a comprehensive model of instruction for English learners that is not only empirically validated, but is appropriate, essential, and *doable* for teachers. Yes, it takes time and effort to create, write, and then teach SIOP® lessons, but the results are well worth it.

Our overarching goal for this book, and for the other SIOP® books in the series, is to help you pull it all together as you create your SIOP® science lesson plans and units. These resources should further your confidence in how to effectively teach science to your English learners. We hope that as you become a successful SIOP® teacher, you will find rewards in your students' growth in both the academic language of science and the scientific knowledge of your course's standards and curriculum.

appendix a: SIOP® Protocol and Component Overview

(Echevarria, Vogt, & Short, 2000; 2004; 2008)

The SIOP® Model was designed to help teachers systematically, consistently, and concurrently teach grade-level academic content and academic language to English learners (ELs). Teachers have found it effective with both ELs and native-English speaking students who are still developing academic literacy. The model consists of eight components and 30 features. The following brief overview is offered to remind you of the preparation and actions teachers should undertake in order to deliver effective SIOP® instruction.

1. Lesson Preparation

The focus for each SIOP® science lesson is the content and language objectives. We suggest that the objectives be linked to curriculum standards and the academic language students need for success in science. Your goal is to help students gain important experience with key grade-level content and skills as they progress toward fluency in English. Hopefully, you now post and discuss the objectives with students each day, even if one period continues a lesson from a previous day, so that students know what they are expected to learn and/or be able to do by the end of that lesson. When you provide a road map at the start of each lesson, students focus on what is important and take an active part in their learning process.

The Lesson Preparation component also advocates for supplementary materials (e.g., visuals, multimedia, adapted or bilingual texts, study guides) because grade-level texts are often difficult for many English learners to comprehend. Graphics or illustrations may be used to make content meaningful—the final feature of the component. It is important to remember that meaningful activities provide access to the key concepts in your science lessons; this is much more important than just providing "fun" activities that students readily enjoy. Certainly, "fun" is good, but "meaningful" and "effective" are better. You will also want to plan tasks and projects for students so they have structured opportunities for oral interaction throughout your lessons.

2. Building Background

In SIOP® lessons, you are expected to connect new concepts with students' personal experiences and past learning. As you prepare ELs for science lessons, you may at times have to build background knowledge because many English learners either have not been studying in U.S. schools or are unfamiliar with American culture. At other times, you may need to activate your students' prior knowledge in order to find out what they already know, to identify misinformation, or to discover when you need to fill in gaps.

The SIOP® Model places importance on building a broad vocabulary base for students. We need to increase vocabulary instruction across the curriculum so our students will become effective readers, writers, speakers, and listeners—and scientists. As a science teacher, you already explicitly teach key vocabulary. Go even further for your ELs by helping them develop word learning strategies such as using context clues, word parts (e.g., affixes), visual aids (e.g., illustrations), and cognates (a word related in meaning and form to a word in another language). Then be sure to design lesson activities that give students multiple opportunities to use new science vocabulary both orally and in writing, such as those found in this book and in *99 Ideas and Activities for Teaching with the SIOP® Model* (Vogt & Echevarria, 2008). In order to move words from receptive knowledge to expressive use, vocabulary needs reinforcement through different learning modes.

3. Comprehensible Input

If you present information in a way that students cannot understand, such as an explanation that is spoken too rapidly, or texts that are far above students' reading levels with no visuals or graphic organizers to assist them, many students—including English learners—will be unable to learn the necessary content. Instead, modify "traditional" instruction with a variety of ESL methods and SIOP® techniques so your students can comprehend the lesson's key concepts. These techniques include, among others:

- Teacher talk appropriate to student proficiency levels (e.g., simple sentences, slower speech)
- Demonstrations and modeling (e.g., modeling how to complete a task or problem)
- Gestures, pantomime, and movement
- Role-plays, improvisation, and simulations
- Visuals, such as pictures, real objects, illustrations, charts, and graphic organizers
- Restatement, paraphrasing, repetition, and written records of key points on the board, transparencies, or chart paper
- Previews and reviews of important information (perhaps in the native language, if possible and as appropriate)
- Hands-on, experiential, and discovery activities

Remember, too, that academic tasks must be explained clearly, both orally and in writing for students. You cannot assume English learners know how to do an assignment because it is a regular routine for the rest of your students. Talk through the procedures and use models and examples of good products and appropriate participation, so students know the steps they should take and can envision the desired result.

When you are dealing with complicated and abstract concepts, it can be particularly difficult to convey information to less proficient students. You can boost the comprehensibility of what you're teaching through native language support, if possible. Supplementary materials (e.g., adapted texts or CDs) in a student's primary language may be used to introduce a new topic, and native language tutoring (if available) can help students check their understanding.

4. Strategies

This component addresses student learning strategies, teacher-scaffolded instruction, and higher-order thinking skills. By explicitly teaching cognitive and metacognitive learning strategies, you help equip students for academic learning both inside and outside the SIOP® classroom. You should capitalize on the cognitive and metacognitive strategies students already use in their first language because those will transfer to the new language.

As a SIOP® teacher, you must frequently scaffold instruction so students can be successful with their academic tasks. You want to support their efforts at their current performance level, but also move them to a higher level of understanding and accomplishment. When students master a skill or task, you remove the supports you provided and add new ones for the next level. Your goal, of course, is for English learners to be able to work independently. They often achieve this independence one step at a time.

You need to ask your ELs a range of questions, some of which should require critical thinking. It is easy to ask simple, factual questions, and sometimes we fall into that trap with beginning English speakers. We must go beyond questions that can be answered with a one- or two-word response, and instead, ask questions and create projects or tasks that require students to think more critically and apply their language skills in a more extended way. Remember this important adage: "Just because ELs don't speak English proficiently doesn't mean they can't *think*."

5. Interaction

We know that students learn through interaction with one another and with their teachers. They need oral language practice to help develop and deepen their content knowledge and support their second language skills. Clearly, you are the main role model for appropriate English usage, word choice, intonation, fluency, and the like, but do not discount the value of student–student interaction. In pairs and small groups, English learners practice new language structures and vocabulary that you have taught as well as important language functions, such as asking for clarification, confirming interpretations, elaborating on one's own or another's idea, and evaluating opinions.

Don't forget that sometimes the interaction patterns expected in an American classroom differ from students' cultural norms and prior schooling experiences. You will want to be sensitive to sociocultural differences and work with students to help them become competent in the culture you have established in your classroom, while respecting their values.

6. Practice & Application

Practice and application of new material is essential for all learners. Our research on the SIOP® Model found that lessons with hands-on, visual, and other kinesthetic tasks benefit ELs because students practice the language and content knowledge through multiple modalities. As a SIOP® science teacher, you want to make sure your lessons include a variety of activities that encourage students to apply both the science content and the English language skills they are learning.

7. Lesson Delivery

If you have delivered a successful SIOP® lesson, that means that the planning you did worked—the content and language objectives were met, the pacing was appropriate, and the students had a high level of engagement. We know that lesson preparation is crucial to effective delivery, but so are classroom management skills. We encourage you to set routines, make sure students know the lesson objectives so they can stay on track, and introduce (and revisit) meaningful activities that appeal to students. Don't waste time, but be mindful of student understanding so that you don't move a lesson too swiftly for students to grasp the key information.

8. Review & Assessment

Each SIOP® science lesson needs time for review and assessment. You will do your English learners a disservice if you spend the last five minutes teaching a new concept rather than reviewing what they have learned so far. Revisit key vocabulary and concepts with your students to wrap up each lesson. Check on student comprehension frequently throughout the lesson period so you know whether additional explanations or reteaching are needed. When you assess students, be sure to provide multiple measures for students to demonstrate their understanding of the content. Assessments should look at the range of language and content development, including measures of vocabulary, comprehension skills, and content concepts.

WHY IS THE SIOP® MODEL NEEDED NOW?

We all are aware of the changing demographics in our U.S. school systems. English learners are the fastest growing subgroup of students and have been for the past two decades. According to the U.S. Department of Education in 2006, English learners numbered 5.4 million in U.S. elementary and secondary schools, about 12% of the student population, and they are expected to comprise about 25% of that population by 2025. In several states, this percentage has already been exceeded. The educational reform movement, and the No Child Left Behind (NCLB) Act in particular, has had a direct impact on English learners. States have implemented standards-based instruction and high-stakes testing, but in many content classes, little or no accommodation is made for the specific language development needs of English learners; this raises a significant barrier to ELs' success because they are expected to achieve high academic standards in English. In many states, ELs are required to pass end-of-grade tests in order to be promoted and/or exit exams in order to graduate.

Unfortunately, teacher development has not kept pace with the EL growth rate. Far too few teachers receive an undergraduate education that includes coursework in English as a second language (ESL) methodologies, which can be applied in content classes through sheltered instruction, and in second language acquisition theory, which can help teachers understand what students should be able to accomplish in a second language according to their proficiency levels, prior schooling, and sociocultural backgrounds. At the end of 2008, only four states—Arizona, California, Florida, and New York—required some undergraduate coursework in these areas for all teacher candidates.

Some teachers receive inservice training in working with ELs from their schools or districts, but it is rarely sufficient for the task they confront. Teachers are expected to teach ELs the new language, English, so the ELs can attain a high degree of proficiency, and in addition, instruct them in all the topics of the different grade-level content courses (more often than not taught in English). A survey conducted by Zehler and colleagues (2003) in 2002 found that approximately 43% of elementary and secondary teachers had ELs in their classrooms, yet only 11% were certified in bilingual education and only 18% in English as a second language. In the five years prior to the survey, teachers who worked with three or more ELs had received on average four hours of inservice training in how to serve them—hardly enough to reach a satisfactory level of confidence and competence.

Even teachers who have received university preparation in teaching English learners report limited opportunities for additional professional development. In a recent survey that sampled teachers in 22 small, medium, and large districts in California, the researchers found that during the previous five years, "forty-three percent of teachers with 50 percent or more English learners in their classrooms had received no more than one in-service that focused on the instruction of English learners" (Gandara, Maxwell-Jolly & Driscoll, 2005, p. 13). Fifty percent of the teachers with somewhat fewer students (26%–50% English learners in their classes) had received either no such inservice or only one. The result of this paucity of professional development is that ELs sit in classes with teachers and other staff who lack expertise in second language acquisition, multicultural awareness, and effective, research-based classroom practices.

It is not surprising, then, that ELs have experienced persistent underachievement on high-stakes tests and other accountability measures. On nearly every state and national assessment, ELs lag behind their native-English speaking peers and demonstrate significant achievement gaps (Kindler, 2002; Kober, et al., 2006; Lee, Grigg & Dion, 2007; Lee, Grigg & Donahue, 2007). In addition to having underqualified teachers, ELs are also more likely to be enrolled in poor, majority-minority schools that have fewer resources and teachers with less experience and fewer credentials than those serving English-proficient students (Cosentino de Cohen, Deterding & Clewell, 2005).

Lower performance on assessments is also the result of education policy. Although research has shown that it takes several years of instruction to become proficient in English (four to nine years, depending on a student's literacy level in the native language and prior schooling) (Collier, 1987; Cummins, 2006; Genesee, Lindholm-Leary, Saunders, & Christian, 2006), current NCLB policy forces schools to test ELs in reading after one year of U.S. schooling in grades 3–8 and one grade in high school. English learners are supposed to take the tests in mathematics and science from the start. Adding to the disconnect between research and policy is the fact that these tests have been designed for native English speakers, rendering them neither valid nor reliable for ELs (AERA, APA, & NCME, 2000). By definition, an English learner is **not** proficient in English; as most of these state assessments are in English, the majority of ELs score poorly on them and are unable to demonstrate their real level of understanding of the subject matter.

Even though it is hard to turn around education policy, teachers, schools, districts, and universities do have opportunities to enact changes in professional development and program design. With this book we hope to help English-language arts teachers grow professionally and develop appropriate skills for working with English learners. There are many approaches and numerous combinations of techniques that can be applied to the delivery of sheltered content instruction. Currently, however, the SIOP® Model is the only scientifically validated model of sheltered instruction for English learners, and it has a growing research base (Center for Applied Linguistics, 2007; Echevarria, Richards, Canges & Francis, 2009; Echevarria & Short, in press; Echevarria, Short & Powers, 2006; Short & Richards, 2008). The SIOP® Model is distinct from other approaches in that it offers a field-tested protocol for systematic lesson planning, delivery, and assessment, making its application for teaching English learners transparent for both preservice candidates preparing to be teachers and practicing teachers engaged in staff development. Further, it provides a framework for organizing the instructional practices essential for sound sheltered content instruction.

THE SIOP® MODEL COMPONENTS AND FEATURES

1) **Lesson Preparation**

1. **Content objectives** clearly defined, displayed and reviewed with students
2. **Language objectives** clearly defined, displayed and reviewed with students
3. **Content concepts** appropriate for age and educational background level of students
4. **Supplementary materials** used to a high degree, making the lesson clear and meaningful (e.g., computer programs, graphs, models, visuals)
5. **Adaptation of content** (e.g., text, assignment) to all levels of student proficiency
6. **Meaningful activities** that integrate lesson concepts (e.g., interviews, letter writing, simulations, models) with language practice opportunities for reading, writing, listening, and/or speaking

2) **Building Background**

7. **Concepts explicitly linked** to students' background experiences
8. **Links explicitly made** between past learning and new concepts

9. **Key vocabulary emphasized** (e.g., introduced, written, repeated, and highlighted for students to see)

3) Comprehensible Input

10. **Speech** appropriate for students' proficiency levels (e.g., slower rate, enunciation, and simple sentence structure for beginners)
11. **Clear explanation** of academic tasks
12. **A variety of techniques** used to make content concepts clear (e.g., modeling, visuals, hands-on activities, demonstrations, gestures, body language)

4) Strategies

13. Ample opportunities provided for students to use **learning strategies**
14. **Scaffolding techniques** consistently used, assisting and supporting student understanding (e.g., think-alouds)
15. A variety of **questions or tasks that promote higher-order thinking skills** (e.g., literal, analytical, and interpretive questions)

5) Interaction

16. Frequent opportunities for **interaction** and discussion between teacher / student and among students, which encourage elaborated responses about lesson concepts
17. **Grouping configurations** support language and content objectives of the lesson
18. Sufficient **wait time for student responses** consistently provided
19. Ample opportunities for students to **clarify key concepts in Ll** as needed with aide, peer, or L1 text

6) Practice & Application

20. **Hands-on materials and / or manipulatives** provided for students to practice using new content knowledge
21. Activities provided for students to **apply content and language knowledge** in the classroom
22. Activities integrate all **language skills** (i.e., reading, writing, listening, and speaking)

7) Lesson Delivery

23. **Content objectives** clearly supported by lesson delivery
24. **Language objectives** clearly supported by lesson delivery
25. **Students engaged** approximately 90% to 100% of the period
26. **Pacing** of the lesson appropriate to students' ability levels

8) Review & Assessment

27. Comprehensive **review of key vocabulary**
28. Comprehensive **review of key content concepts**
29. Regular **feedback** provided to students on their output (e.g., language, content, work)
30. **Assessment of student comprehension and learning** of all lesson objectives (e.g., spot checking, group response) throughout the lesson

appendix b: Academic Science Vocabulary Based on Sample State Standards

Grade Band: K–2

Life Science	*Earth Science*	*Physical Science*	*Investigation & Experimentation*	*Measurement*	*Data Presentation*	*General Academic*
amphibians	cloud	attract	classify	area	bar graph	ask
animal	deserts	color	collect	calendar	data	because
behavior	Earth	compass	conduct	clock	display	characteristic
birds	energy	force	construct	day	graph	communicate
environment	evaporation	fossils	equipment	distance	lists	demonstration
feathers	landforms	gas	estimate	foot	picture graph	description
fin	moon	gravity	evidence	height	pie chart	different
flower	mountains	hardness	experiment	hour	table	discuss
fruit	oceans	liquid	hypothesis	inch	timeline	explain
growth	planets	magnets	instructions	length		feature
habitat	precipitation	matter	investigate	measure		label
insects	rain	minerals	laboratory	minute		model
leaf	resources	motion	observe	ounce		object
life cycle	rivers	poles	plan	pound		organize
living	seasons	property	predict	ruler		pattern
mammal	show	repel	record	second		position
non-living	soil	rocks	result	senses		question
plant	stars	shape	select	size		relationship
recycle	sun	solid	solve	temperature		report
reptiles	thermometer	substance	sort	unit of measurement		sequence
root	valleys	texture		volume		similar
scales	water cycle			week		technology
seeds	weather			weight		
stem	wind			width		
tail				year		
wing						

Grade Band: 3–5 (adds to K–2 Vocabulary)

Life Science	*Earth Science*	*Physical Science*	*Investigation & Experimentation*	*Measurement*	*Data Analysis & Representation*	*General Academic*
adaptation	atmosphere	atom	analysis	balance	chart	application
aquatic	bedrock	buoyant force	claim	capacity	circle graph	argument
camouflage	climate	compound machine	conclusion	Celsius	data collection method	cause
carnivore	condensation	conductivity	confirm	centimeter	diagram	chronological
circulatory system	earthquake	electric current	cover slip	circumference	key	comparison
community	elevation	electromagnetic	criteria	depth	line graph	component
consumer	erosion	element	discovery	elapsed time	map	distinguish
decomposer	eruption	frequency	infer	English system of measurement	plot	effect
digestive system	fossil fuel	insulation	inquiry	Fahrenheit	sample	function
diversity	geology	magnification	method	graduated cylinder	survey	if–then
ecosystem	land surface	metal	microscope	gram	tallies	impact
excretory system	lunar	molecule	outcome	liter		influence
food chain	magma	opaque	procedure	mass		justify
food web	nonrenewable	parallel circuit	process	meter		logical
herbivore	nutrient	prism	product	metric system		opinion
hibernation	orbit	reflect	reaction	standard vs. nonstandard units		permanent
instinct	renewable	refract	slide	surface area		perspective
interdependency	revolve	series circuit	test	unit conversion		problem-solving
invertebrate	rotate	simple machine	trial			quantity
migration	solar system	sound waves	variable			resemble
mimicry	sources of energy	spectrum				temporary
nervous system	telescope	translucent				
omnivore	topsoil	transparent				

Life Science	*Earth Science*	*Physical Science*	*Investigation & Experimentation*	*Measurement*	*Data Analysis & Representation*	*General Academic*
organism	volcano	vibrate				
photosynthesis	water vapor	visible light				
pollination	weathering	wavelength				
population						
predator						
prey						
producer						
reproduce						
respiration						
species						
survival						
terrestrial						
vertebrate						

Grade Band: 6–8 (adds to K–5 Vocabulary)

Life Science	*Earth Science*	*Physical Science*	*Investigation & Experimentation*	*Measurement*	*Data Analysis & Representation*	*General Academic*
abiotic	constellations	acids	accuracy	computer-linked probe	alternative explanations	attribute
asexual reproduction	deposit	activation energy	approximate	electron microscope	database	as a result
biome	epicenter	alternating circuit	conditions	reference point	dependent variable	catastrophic
biotic	fault	amperes	constant	scale	experimental error	challenge
carbon cycle	galaxy	angle of incidence	control group	slope	frequency distribution	contrary to
cell	geologic layers	angle of reflection	controlled tests	triple beam balance	generalization	differentiate
cell division	igneous	atomic mass	data set		graphical representation	distinguish
cell theory	light year	atomic number	devise		independent variable	dynamic
cellular transport	mantle	balanced equations	inferences		interpretation	evaluate
chromosomes	metamorphic	bases	manipulate		linear	except
climate change	petroleum	catalyst	parameters		mean	external
competition	plate tectonics	chemical bonds	precise		nonlinear	formulate
cooperation	radioactive dating	chemical change	random sample		qualitative	however
DNA	rock cycle	chemical compounds	repeated trials		quantitative	in conclusion
dominant	sediment	chemical reactions	reproducibility		range	internal
ecology	sedimentary	combustibility	sample size		scale model	phenomenon
extinction	solar energy	concentration	treatment group		scatterplot	practical
fertilization	space exploration	conduction			scientific reasoning	reject
food pyramid	topography	convection			spreadsheet	static
genes		covalent bond			validity	therefore
genetic engineering		density			variable	transmit
genetics		direct circuit				unless
host		electrons				
inherit		energy consumption				

Life Science	*Earth Science*	*Physical Science*	*Investigation & Experimentation*	*Measurement*	*Data Analysis & Representation*	*General Academic*
locomotion		energy transformations				
meiosis		fission				
mitosis		friction				
mutation		fusion				
natural selection		gravitational force				
nitrogen cycle		inorganic				
nourishment		ionic bond				
offspring		kinetic energy				
organelle		neutralization				
parasite		neutrons				
phototropism		nuclear reaction				
recessive		organic				
sexual reproduction		oxidation				
social hierarchy		Periodic table of elements				
symbiosis		pH scale				
		physical change				
		potential energy				
		protons				
		solubility				
		solute				
		solvent				
		speed				
		valence electrons				
		velocity				
		voltage				
		wave theory				

Grade Band: 9–12 (adds to K–8 Vocabulary)

Earth Science	Biology	Chemistry	Physics	Data Analysis & Representation	General Academic
aquifer	allele	aldehydes	acceleration	causation	analogy
astronomy	antibodies	alkaline	average speed	correlation	cumulative
big bang model	ATP	amines	balanced force	frequency counts	derive
biomass	biochemical reaction	chromatography	capacitor	graphing calculator	fluctuation
black hole	encode	diffusion	constant speed	independent trials	fraudulent
celestial	eukaryote	distillation	curvilinear motion	inferential statistics	inconsistent
cosmic radiation	gamete	electronegativity	Doppler effect	normal curve	paradigm
cosmology	gene pool	electrostatic attraction	entropy	probability	principle
energy reservoir	genetic drift	endothermic	force field	random sampling technique	propagate
energy sink	genotype	equilibrium	magnitude	representativeness of sample	reliability
gas planet	growth curve	esters	oscillation	sampling distribution	resolve
gravitational collapse	heterozygous	ethers	plasma	standard deviation	skeptical
greenhouse effect	homeostasis	exothermic	polarization	statistic	spontaneous
hydrologic cycle	homozygous	half-life	quantum	statistical significance	stabilize
light years	hormones	halogen	relativity	time intervals	synthesis
nonrenewable resource	immune response	ideal gas	resistor	upper/lower bounds	uncertainty
ozone	limiting factor	inert	robotics	variance	uniformly
renewable resource	lineage	ionization energy	thermodynamics		verify
satellite imagery	metabolic	isotope	trajectory		
seismic waves	neuron	Kelvin temperature	transistor		
star chart	phenotype	ketones	unbalanced force		
temperature inversions	prokaryote	Lewis dot structure	vacuum		
terrestrial planet	recombination	molar mass	vector		

Earth Science	*Biology*	*Chemistry*	*Physics*	*Data Analysis & Representation*	*General Academic*
ultraviolet radiation	RNA	mole			
	specialization	oxidation-reduction reaction			
	succession factor	photon			
	transcription	polymer			
	translation	polypeptide			
	virus	quark			
	X-linked	radiation			
	zygote	radioactive decay			
		stoichiometry			

appendix c: Blackline Masters

1 t-chart: objects in the sky

Name ______________________ Date ______________ Period ______________

Objects That Give Light	Objects That Do Not Give Light
sun	clouds

Name: ____________________

Sun and Moon Phases

sunrise	midday	sunset
moonrise	midnight	moonset

Sun and Moon Phases—Side 2

sunrise	midday	sunset
I wake up. It is cold outside. My mom drinks coffee.		
moonrise	midnight	moonset

An object at rest stays at rest
AND
an object in motion stays in motion
UNLESS
a force acts upon it.

Running Observations:

-
-
-
-

Flick the Note Card Observations:

-
-
-
-

Pile of Pennies Observations:

-
-
-
-

Now, write a complete sentence for each observation using the following sentence starter.:

One observation I made during the experiment was . . .

Running:

Flick the Note Card:

Pile of Pennies:

Forces Are Fun! Observations

Record observations and complete the sentence as you conduct the experiments:

Bucket of Water

•

•

•

One interesting thing I noticed was____

____

Magnets and Metal

•

•

•

One interesting thing I noticed was____

____

Forces in Water

•

•

•

One interesting thing I noticed was____

____

Paper Drop

•

•

•

One interesting thing I noticed was____

____

Hot Hands

•

•

•

One interesting thing I noticed was____

____

1. Record the type of force demonstrated in the experiments.

Experiment	Bucket of Water	Magnets and Metal	Forces in Water	Paper Drop	Hot Hands
Force	*centripetal*				

2. Write summary statements about Mini-Experiments #2–#5, using the sentence frame:

"____________________ was and example of ____________________."

Bucket of Water

The water staying in the bucket was an example of centripetal force.

Magnets and Metal

Forces in Water

Paper Drop

Hot Hands

6 raisin lab report

Name ______________________ Date ____________________ Period ____________

The Mysterious Raisins

In the space below, record your **observations** of the raisins in the beaker.

*

*

*

*

*

*

*

*

*

In the space below, use the words **buoyant force** and **gravity** to describe what you saw happening in the beaker. Write at least two sentences.

____

____

____

raisin lab report

In the space below, use Newton's First Law of Motion to describe what you saw in the beaker. Write at least two sentences.

An object at rest stays at rest
AND
an object in motion stays in motion
UNLESS
a force acts upon it.

Observation	Explanation of Newton's Law

7 everyday forces chart

real life object	rest or motion	changes to motion or rest	force acting on the object	type of force (gravity, centripetal, magnetic, buoyant)	What happened?

One example of Newton's Law in everyday life is ______________________________

When the ______________________________ is at rest, it stays at rest because ______________________________

When the ______________________________ is in motion, it stays in motion because ______________________________

When the ______________________________ is at rest, it becomes in motion because ______________________________

When the ______________________________ is in motion, it becomes at rest because ______________________________

Friction affects the ______________________________ and ______________________________

Gravity affects the ______________________________ and ______________________________

Centripetal force affects the ______________________________ and ______________________________

Buoyant force affects the ______________________________ and ______________________________

Writer's Checklist

- ☐ I identified an everyday example of Newton's First Law of Motion.
- ☐ I described how the object moves from rest to motion or motion to rest.
- ☐ I identified the force that acts on the object.
- ☐ I identified the type of force that acts on the object.
- ☐ I used at least one key vocabulary word from each column on the chart.
- ☐ I used at least 5 complete sentences. (I might decide to use some of the sentence starters.)

Rock Cycle

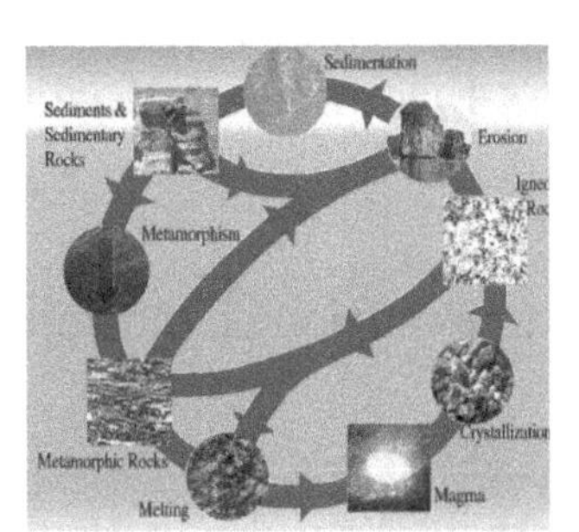

Fine Grain

Hardness

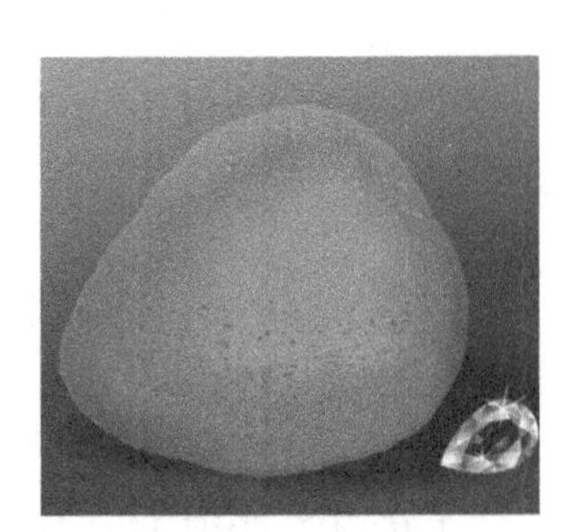

Texture

Coarse Grain

Color

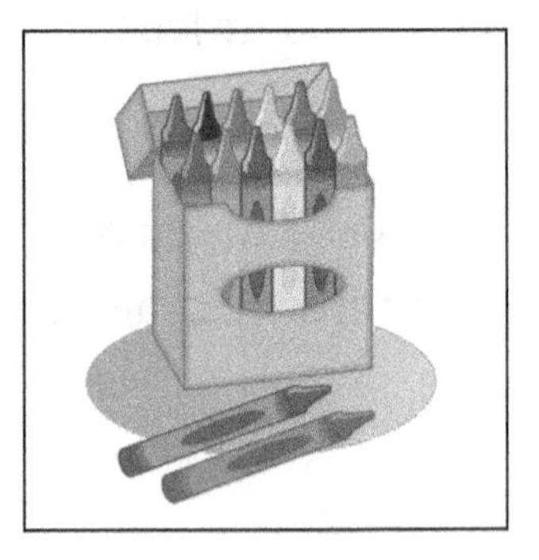

Igneous Rock

Metamorphic Rock

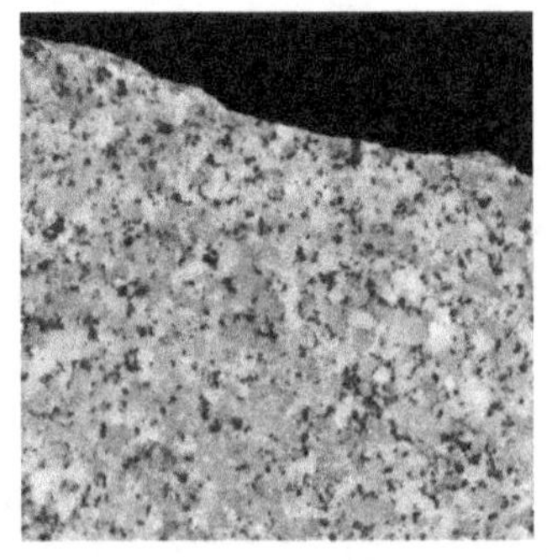

Sedimentary Rock

(continued)

Crystals

Pressure

Follated

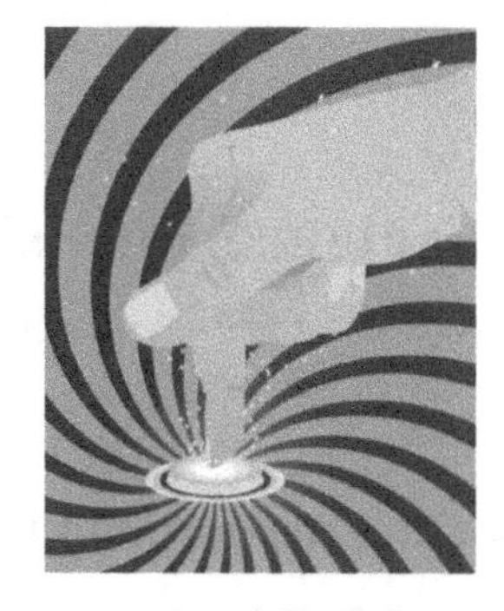

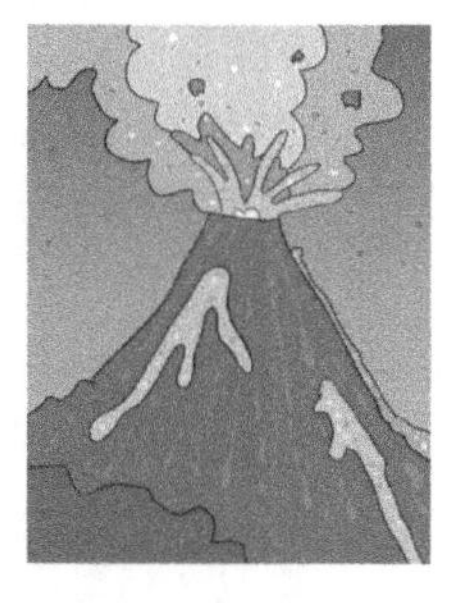

Extrusive Igneous Rock

Intrusive Igneous Rock

Porous

Clastic

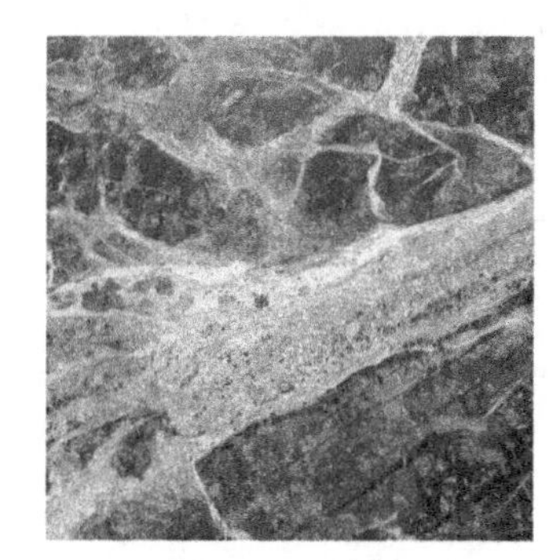

Name ______________________________ Date ____________ Period ____________

Science Rocks!

Directions: In the box, write your observations about the rocks at your table.

*

*

*

*

*

Use the space below to write your sentence.

One observation I made of the rocks is

____

____

____

____

1

(continued)

Name ______________________ Date ____________ Period ____________

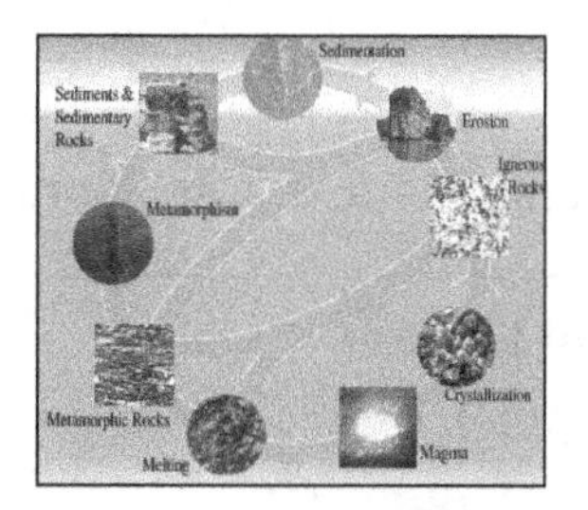

Science Rocks!

Directions: Record observations of each type of rock. Write in complete sentences.

Sedimentary Rocks	Igneous Rocks	Metamorphic Rocks
*	*	*
*	*	*
*	*	*
*	*	*
*	*	*

2

(continued)

Name ______________________ Date ____________ Period ____________

Directions: Record facts about this type of rock from your text. Use your own words.

Sedimentary Rocks 
*   *   *   *   *   *

3

(continued)

Name ______________________ Date ____________ Period ____________

Directions: Record facts about this type of rock from your text. Use your own words.

Igneous Rocks

*

*

*

*

*

*

4

(continued)

Name ________________________________ Date _____________ Period _____________

Directions: Record facts about this type of rock from your text. Use your own words.

Metamorphic Rocks 
*   *   *   *   *   *

5

[word]	[illustration]
[definition]	[word in a sentence]

Rock Cycle Letter Rubric

Name ______________________________ Date __________ Period _____

Pretend that you are a rock in the rock cycle. Write a letter to your family telling about your journey through the cycle. Include all of the key vocabulary and use sequence words in your letter. Use the rubric below to guide you in your planning and implementation of the project.

Requirements

Student correctly uses each vocabulary word ____________/13 points

rock cycle

metamorphism

melting

crystallization

weathering

transportation

deposition

compaction

cementation

pressure

igneous

metamorphic

sedimentary

Student uses at least five sequence words ____________/5 points

Student includes the four sections of a friendly letter____________/4 points

(date, introduction, body, closing)

Student puts best effort into the assignment ____________/3 points

TOTAL ____________/25 points

Cheek Cell and Onion Skin Cell Lab

Name __________________________________ Date ____________ Period __________

In the space below, draw a diagram of what you see in the microscope when you look at the cheek cells.

In the space below, write three detailed scientific observations of what you see.

*

*

*

(continued)

13 cell lab report

In the space below, draw a diagram of what you see in the microscope when you look at the onion skin cells.

In the space below, write three detailed scientific observations of what you see.

*

*

*

After the class discussion, go back to your drawings and label the cell wall, cell membrane, cytoplasm, and nucleus.

Ribosomes are small organelles that make protein for the cell.

Golgi complex (also called *Golgi apparatus* or *Golgi body*) is an organelle found in the cytoplasm that stores and moves nutrients through the cell.

Cytoplasm is a jellylike substance that is found within the cell that holds all of the cell's organelles.

The **nucleus** is the "control center" of the eukaryotic cell. The nucleus contains the genetic material, which is called DNA.

Nucleolus is a structure found in the nucleus that makes ribosomes.

The **nuclear membrane** is a membrane that surrounds the nucleus and controls what can go into and out of the nucleus.

The **cell membrane** (also known as the *plasma membrane*) is a membrane around the cell that helps to control what can go into and out of the cell.

A **mitochondrion** (the plural is *mitochondria*) is an organelle that makes energy for the cell.

(continued)

14 organelles definition strips

Rough endoplasmic reticulum (rough ER) is the endoplasmic reticulum that contains ribosomes and is responsible for making protein.

Smooth endoplasmic reticulum (smooth ER) is the endoplasmic reticulum that does not have ribosomes and is responsible for making fats.

A **vacuole** is a sack of stuff inside the cell that helps to give the cell structure and support. In plant cells, vacuoles are relatively big and in animal cells, vacuoles are small.

The **cell wall** is a thick layer of fibers that is found around the outside of the cell membrane in plant cells. The cell wall gives support to the cell.

A **chloroplast** is an organelle found only in plant cells. A chloroplast is the place where the plant makes its food.

A **lysosome** is a sack of enzymes found inside the cytoplasm.

A **centriole** is a small structure in an animal cell that helps during cell division.

Cell Organelle Structure and Function

Name ______________________________ Date __________ Period _____

Name of Organelle	Structure What does it look like? What is it made of?	Function What does it do? What is its purpose?

Name ______________________ Date ____________ Period ____________

Snowball!

____
Name 2

____
Name 3

Signal Words
If you are asked to Compare, use these words: both alike the same as similar to also neither in comparison
If you are asked to Contrast, use these words: one but not the other different from differs unlike in contrast however

The Cell Project

Name ______________________________ Date __________ Period ______

You will create a project that shows your understanding of eukaryotic cells. Your project will be creative and full of information about plant and animal cells. Use the rubric below to guide you in your planning and implementation of the project.

Requirements

The project has a title. ________/2

The project has proper spelling and grammar. ________/4

The project is organized, neat, and easy-to-follow. ________/3

The project contains information about the **structure** *and* **function** of the following organelles and **uses each word properly in context**. ______/33

Cell wall	________/3
Cell membrane	________/3
Cytoplasm	________/3
Mitochondria	________/3
Golgi body	________/3
Rough endoplasmic reticulum	________/3
Smooth endoplasmic reticulum	________/3
Ribosomes	________/3
Nucleus	________/3
Chloroplasts	________/3
Vacuoles	________/3

There is information about where each organelle is found (animal cells, plant cells, or both). ________/3

You put your best effort into this project. ________/5

TOTAL ________**/50**

appendix d: The Sheltered Instruction Observation Protocol (SIOP®)

Observer(s): ______________________ Teacher: ______________________
Date: ______________________ School: ______________________
Grade: ______________________ Class/Topic: ______________________
ESL Level: ______________________ Lesson: Multi-day Single-day *(circle one)*

Total Points Possible: 120 (Subtract 4 points for each NA given: ________)
Total Points Earned: ________ Percentage Score: ________

Directions: Circle the number that best reflects what you observe in a sheltered lesson. You may give a score from 0–4 (or NA on selected items). Cite under "Comments" specific examples of the behaviors observed.

LESSON PREPARATION

4	3	2	1	0
1. **Content objectives** clearly defined, displayed and reviewed with students		**Content objectives** for students implied		No clearly defined **content objectives** for students

Comments:

4	3	2	1	0
2. **Language objectives** clearly defined, displayed and reviewed with students		**Language objectives** for students implied		No clearly defined **language objectives** for students

Comments:

4	3	2	1	0
3. **Content concepts** appropriate for age and educational background level of students		**Content concepts** somewhat appropriate for age and educational background level of students		**Content concepts** inappropriate for age and educational background level of students

Comments:

4	3	2	1	0
4. **Supplementary materials** used to a high degree, making the lesson clear and meaningful (e.g., computer programs, graphs, models, visuals)		Some use of **supplementary materials**		No use of **supplementary materials**

Comments:

(Echevarria, Vogt, & Short, 2000; 2004; 2008)

4	3	2	1	0	NA
5. **Adaptation of content** (e.g., text, assignment) to all levels of student proficiency		Some **adaptation of content** to all levels of student proficiency		No significant **adaptation of content** to all levels of student proficiency	

Comments:

4	3	2	1	0
6. **Meaningful activities** that integrate lesson concepts (e.g., interviews, letter writing, simulations, models) with language practice opportunities for reading, writing, listening, and/or speaking		**Meaningful activities** that integrate lesson concepts but provide few language practice opportunities for reading, writing, listening, and/or speaking		No **meaningful activities** that integrate lesson concepts with language practice

Comments:

BUILDING BACKGROUND

4	3	2	1	0	NA
7. **Concepts explicitly linked** to students' background experiences		**Concepts loosely linked** to students' background experiences		**Concepts not explicitly linked** to students' background experiences	

Comments:

4	3	2	1	0
8. **Links explicitly made** between past learning and new concepts		**Few links made** between past learning and new concepts		**No links made** between past learning and new concepts

Comments:

4	3	2	1	0
9. **Key vocabulary** emphasized (e.g., introduced, written, repeated, and highlighted for students to see)		**Key vocabulary** introduced, but not emphasized		**Key vocabulary** not introduced or emphasized

Comments:

COMPREHENSIBLE INPUT

4	3	2	1	0
10. **Speech** appropriate for students' proficiency levels (e.g., slower rate, enunciation, and simple sentence structure for beginners)		**Speech** sometimes inappropriate for students' proficiency levels		**Speech** inappropriate for students' proficiency levels

Comments:

4	3	2	1	0
11. **Clear explanation** of academic tasks		**Unclear** explanation of academic tasks		**No** explanation of academic tasks

Comments:

4	3	2	1	0
12. **A variety of techniques** used to make content concepts clear (e.g., modeling, visuals, hands-on activities, demonstrations, gestures, body language)		Some **techniques** used to make content concepts clear		No **techniques** used to make concepts clear

Comments:

STRATEGIES

4	3	2	1	0
13. Ample opportunities provided for students to use **learning strategies**		Inadequate opportunities provided for students to use **learning strategies**		No opportunity provided for students to use **learning strategies**

Comments:

4	3	2	1	0
14. **Scaffolding techniques** consistently used, assisting and supporting student understanding (e.g., think-alouds)		**Scaffolding techniques** occasionally used		**Scaffolding techniques** not used

Comments:

4	3	2	1	0
15. A variety of **questions or tasks that promote higher-order thinking skills** (e.g., literal, analytical, and interpretive questions)		Infrequent **questions or tasks that promote higher-order thinking skills**		No **questions or tasks that promote higher-order thinking skills**

Comments:

INTERACTION

4	3	2	1	0
16. Frequent opportunities for **interaction** and discussion between teacher/student and among students, which encourage elaborated responses about lesson concepts		**Interaction** mostly teacher-dominated with some opportunities for students to talk about or question lesson concepts		**Interaction** teacher-dominated with no opportunities for students to discuss lesson concepts

Comments:

4	3	2	1	0
17. **Grouping configurations** support language and content objectives of the lesson		**Grouping configurations** unevenly support the language and content objectives		**Grouping configurations** do not support the language and content objectives

Comments:

4	3	2	1	0
18. Sufficient **wait time for student responses** consistently provided		Sufficient **wait time for student responses** occasionally provided		Sufficient **wait time for student responses** not provided

Comments:

4	3	2	1	0	NA
19. Ample opportunities for students to **clarify key concepts in L1** as needed with aide, peer, or L1 text		Some opportunities for students to **clarify key concepts in L1**		No opportunities for students to **clarify key concepts in L1**	

Comments:

PRACTICE & APPLICATION

4	3	2	1	0	NA
20. **Hands-on materials and/or manipulatives** provided for students to practice using new content knowledge		Few **hands-on materials and/or manipulatives** provided for students to practice using new content knowledge		No **hands-on materials and/or manipulatives** provided for students to practice using new content knowledge	

Comments:

4	3	2	1	0	NA
21. Activities provided for students to **apply content and language knowledge** in the classroom		Activities provided for students to **apply** either **content or language knowledge** in the classroom		No activities provided for students to **apply content and language knowledge** in the classroom	

Comments:

4	3	2	1	0
22. Activities integrate all **language skills** (i.e., reading, writing, listening, and speaking)		Activities integrate some **language skills**		Activities do not integrate **language skills**

Comments:

LESSON DELIVERY

4	3	2	1	0
23. **Content objectives** clearly supported by lesson delivery		**Content objectives** somewhat supported by lesson delivery		**Content objectives** not supported by lesson delivery

Comments:

4	3	2	1	0
24. **Language objectives** clearly supported by lesson delivery		**Language objectives** somewhat supported by lesson delivery		**Language objectives** not supported by lesson delivery

Comments:

4	3	2	1	0
25. **Students engaged** approximately 90% to 100% of the period		**Students engaged** approximately 70% of the period		**Students engaged** less than 50% of the period

Comments:

4	3	2	1	0
26. **Pacing** of the lesson appropriate to students' ability levels		**Pacing** generally appropriate, but at times too fast or too slow		**Pacing** inappropriate to students' ability levels

Comments:

REVIEW & ASSESSMENT

4	3	2	1	0
27. Comprehensive **review of key vocabulary**		Uneven **review of key vocabulary**		No **review of key vocabulary**

Comments:

4	3	2	1	0
28. Comprehensive **review of key content concepts**		Uneven **review of key content concepts**		No **review of key content concepts**

Comments:

4	3	2	1	0
29. Regular **feedback** provided to students on their output (e.g., language, content, work)		Inconsistent **feedback** provided to students on their output		No **feedback** provided to students on their output

Comments:

4	3	2	1	0
30. **Assessment of student comprehension and learning** of all lesson objectives (e.g., spot checking, group response) throughout the lesson		**Assessment of student comprehension and learning** of some lesson objectives		No **assessment of student comprehension and learning** of lesson objectives

Comments:

The Sheltered Instruction Observation Protocol (SIOP®)

(Echevarria, Vogt, & Short, 2000; 2004; 2008)

Observer(s): ______________ Teacher: ______________
Date: ______________ School: ______________
Grade: ______________ Class/Topic: ______________
ESL Level: ______________ Lesson: Multi-day Single-day *(circle one)*

Total Points Possible: 120 (Subtract 4 points for each NA given) ________
Total Points Earned: _____ Percentage Score: _____

Directions: Circle the number that best reflects what you observe in a sheltered lesson. You may give a score from 0–4 (or NA on selected items). Cite under "Comments" specific examples of the behaviors observed.

	Highly Evident		Somewhat Evident		Not Evident	
Preparation	4	3	2	1	0	
1. **Content objectives** clearly defined, displayed, and reviewed with students	❑	❑	❑	❑	❑	
2. **Language objectives** clearly defined, displayed, and reviewed with students	❑	❑	❑	❑	❑	
3. **Content concepts** appropriate for age and educational background level of students	❑	❑	❑	❑	❑	
4. **Supplementary materials** used to a high degree, making the lesson clear and meaningful (e.g., computer programs, graphs, models, visuals)	❑	❑	❑	❑	❑	**NA**
5. **Adaptation of content** (e.g., text, assignment) to all levels of student proficiency	❑	❑	❑	❑	❑	❑
6. **Meaningful activities** that integrate lesson concepts (e.g., surveys, letter writing, simulations, constructing models) with language practice opportunities for reading, writing, listening, and/or speaking	❑	❑	❑	❑	❑	

Comments:

Building Background	4	3	2	1	0	NA
7. **Concepts explicitly linked** to students' background experiences	❑	❑	❑	❑	❑	❑
8. **Links explicitly made** between past learning and new concepts	❑	❑	❑	❑	❑	
9. **Key vocabulary** emphasized (e.g., introduced, written, repeated, and highlighted for students to see)	❑	❑	❑	❑	❑	

Comments:

Comprehensible Input	4	3	2	1	0
10. **Speech** appropriate for students' proficiency levels (e.g., slower rate, enunciation, and simple sentence structure for beginners)	❑	❑	❑	❑	❑
11. **Clear explanation** of academic tasks	❑	❑	❑	❑	❑
12. **A variety of techniques** used to make content concepts clear (e.g., modeling, visuals, hands-on activities, demonstrations, gestures, body language)	❑	❑	❑	❑	❑

Comments:

Strategies	4	3	2	1	0
13. Ample opportunities provided for students to use **learning strategies**	❑	❑	❑	❑	❑

	Highly Evident 4	3	Somewhat Evident 2	1	Not Evident 0	
14. **Scaffolding techniques** consistently used assisting and supporting student understanding (e.g., think-alouds)	❑	❑	❑	❑	❑	
15. A variety of **questions or tasks that promote higher-order thinking skills** (e.g., literal, analytical, and interpretive questions)	❑	❑	❑	❑	❑	
Comments:						
Interaction	4	3	2	1	0	
16. Frequent opportunities for **interaction** and discussion between teacher/student and among students, which encourage elaborated responses about lesson concepts	❑	❑	❑	❑	❑	
17. **Grouping configurations** support language and content objectives of the lesson	❑	❑	❑	❑	❑	
18. Sufficient **wait time for student responses** consistently provided	❑	❑	❑	❑	❑	NA
19. Ample opportunities for students to **clarify key concepts in L1** as needed with aide, peer, or L1 text	❑	❑	❑	❑	❑	❑
Comments:						
Practice & Application	4	3	2	1	0	NA
20. **Hands-on materials and/or manipulatives** provided for students to practice using new content knowledge	❑	❑	❑	❑	❑	❑
21. Activities provided for students to **apply content and language knowledge** in the classroom	❑	❑	❑	❑	❑	❑
22. Activities integrate all **language skills** (i.e., reading, writing, listening, and speaking)	❑	❑	❑	❑	❑	
Comments:						
Lesson Delivery	4	3	2	1	0	
23. **Content objectives** clearly supported by lesson delivery	❑	❑	❑	❑	❑	
24. **Language objectives** clearly supported by lesson delivery	❑	❑	❑	❑	❑	
25. **Students engaged** approximately 90% to 100% of the period	❑	❑	❑	❑	❑	
26. **Pacing** of the lesson appropriate to students' ability level	❑	❑	❑	❑	❑	
Comments:						
Review & Assessment	4	3	2	1	0	
27. Comprehensive **review of key vocabulary**	❑	❑	❑	❑	❑	
28. Comprehensive **review of key content concepts**	❑	❑	❑	❑	❑	
29. Regular **feedback** provided to students on their output (e.g., language, content, work)	❑	❑	❑	❑	❑	
30. **Assessment of student comprehension and learning** of all lesson objectives (e.g., spot checking, group response) throughout the lesson	❑	❑	❑	❑	❑	
Comments:						

references

American Educational Research Association (AERA), American Psychological Association (APA), & National Council on Measurement in Education (NCME). (2000). Position statement of the American Educational Research Association concerning high-stakes testing in pre-K–12 education. *Educational Researcher, 29,* 24–25.

August, D., & Shanahan, T. (Eds.). (2006). *Developing literacy in second-language learners: A report of the National Literacy Panel on Language-Minority Children and Youth.* Mahwah, NJ: Erlbaum.

Aukerman, M. (2007). A culpable CALP: Rethinking the conversational/academic proficiency distinction in early literacy instruction. *The Reading Teachers, 60*(7), 626–635.

Bailey, A. L. (Ed.). (2007). *The language demands of school: Putting academic English to the test.* New Haven, CT: Yale University Press.

Bailey, A., & Butler, F. (2007). A conceptual framework of academic English language for broad application to education. In A. Bailey (Ed.), *The language demands of school: Putting academic English to the test* (pp. 68–102). New Haven, CT: Yale University Press.

Bartolomé, L. I. (1998). *The misteaching of academic discourses: The politics of language in the classroom.* Boulder, CO: Westview Press.

Baumann, J., Jones, L., & Seifert-Kessell, N. (1993). Using think-alouds to enhance children's comprehension monitoring abilities. *The Reading Teacher, 47*(3), 184–193.

Biancarosa, G., & Snow, C. (2004). *Reading next: A vision for action and research in middle and high school literacy.* Report to the Carnegie Corporation of New York. Washington, DC: Alliance for Excellent Education.

Bredderman, T. (1983). "Effects of Activity-Based Elementary Science on Student Outcomes: A Quantitative Synthesis." *Review of Educational Research, 53*(4), 499–518.

Brown, B., & Ryoo, K. (2008). Teaching science as a language: A 'content-first' approach to science teaching. *Journal of Research in Science Teaching, 45*(5), 529–553.

Buehl, D. (2009). *Strategies for interactive learning* (3rd ed.). Newark, DE: International Reading Association.

California Department of Education. (2008). Statewide Stanford 9 test results for reading: Number of students tested and percent scoring at or above the 50th percentile ranking. Retrieved April 1, 2009, from http://www.cde.ca.gov/dataquest/.

California Department of Education. (1998). *Science content standards for California public schools.* Sacramento: Department of Education. Retrieved from cde.ca.gov/be/st/ss/documents/sciencestnds.pdf. December, 2008.

Carr, J., Sexton, U., & Lagunoff, R. (2007). *Making science accessible for English learners: A guidebook for teachers.* San Francisco, CA: WestEd.

Castillo, M. (2008). *Reviewing objectives with English language learners.* Presented at SEI Seminar, Phoenix, AZ.

Cazden, C. (1976). How knowledge about language helps the classroom teacher—or does it? A personal account. *The Urban Review, 9*, 74–91.

Cazden, C. (1986). Classroom discourse. In M. D. Wittrock (Ed.), *Handbook of research on teaching* (3rd ed.) (pp. 432–463). New York, NY: Macmillan.

Cazden, C. (2001). *Classroom discourse: The language of teaching and learning* (2nd. ed.). Portsmouth, NH: Heinemann.

Center for Applied Linguistics. (2007). *Academic literacy through sheltered instruction for secondary English language learners.* Final Report to the Carnegie Corporation of New York. Washington, DC: Center for Applied Linguistics.

Chamot, A. U., & O'Malley, J. M. (1994). *The CALLA handbook: Implementing the cognitive academic language learning approach.* Reading, MA: Addison-Wesley.

Cloud, N., Genesee, F., & Hamayan, E. (2009). *Literacy instruction for English language learners.* Portsmouth, NH: Heinemann.

Collier, V. (1987). Age and rate of acquisition of second language for academic purposes. *TESOL Quarterly, 21*(3), 617–641.

Cosentino de Cohen, C., Deterding, N., & Clewell, B. C. (2005). *Who's left behind? Immigrant children in high and low LEP schools.* Washington, DC: Urban Institute. Retrieved January 2, 2009 at http://www.urban.org/UploadedPDF/411231_whos_left_behind.pdf

Coxhead, A. (2000). A new academic word list. *TESOL Quarterly, 34*(2), 213–238.

Cummins, J. (1979). *Cognitive/academic language proficiency, linguistic interdependence, the optimum age questions, and some other matters.* Working Papers on Bilingualism, No. 19, 121–129. Toronto: Ontario Institute for Studies in Education.

Cummins, J. (2000). *Language, power, and pedagogy: Bilingual children in the crossfire.* Clevedon, UK: Multilingual Matters.

Cummins, J. (2006). How long does it take for an English language learner to become proficient in a second language? In E. Hamayan & R. Freeman (Eds.), *English language learners at school: A guide for administrators* (pp. 59–61). Philadelphia, PA: Caslon Publishing.

Deussen, T., Autio, E., Miller, B., Lockwood, A. T., & Stewart, V. (2008). *What teachers should know about instruction for English language learners.* Portland, OR: Northwest Regional Educational Laboratory.

Dutro, S., & Moran, C. (2003) Rethinking English language instruction: An architectural approach. In G. Garcia (Ed.), *English learners: Reaching the highest level of English literacy* (pp. 227–258). Newark, NJ: International Reading Association.

Echevarria, J. (1995). Interactive reading instruction: A comparison of proximal and distal effects of instructional conversations. *Exceptional Children, 61*(6), 536–552.

Echevarria, J., & Colburn, A. (2006). Designing lessons: Inquiry approach to science using the SIOP® Model. In A. Fathman & D. Crowther (Eds.), *Science for English language learners* (pp. 95–108). Arlington, VA: National Science Teachers Association.

Echevarria, J., & Graves, A. (2007). *Sheltered content instruction: Teaching English language learners with diverse abilities* (3rd ed.). Boston, MA: Allyn & Bacon.

Echevarria, J., Richards, C., Canges, R., & Francis, D. (2009). *Using the SIOP® Model to promote the acquisition of language and science concepts.* Submitted for publication.

Echevarria, J., Short, D., & Powers, K. (2006). School reform and standards-based education: An instructional model for English language learners. *Journal of Educational Research, 99*(4), 195–211.

Echevarria, J., & Silver, J. (1995). *Instructional conversations: Understanding through discussion*. [Videotape]. National Center for Research on Cultural Diversity and Second Language Learning.

Echevarria, J., & Short, D. J. (in press). Programs and practices for effective sheltered content instruction. In D. Dolson & L. Burnham-Massey (Eds.), *Improving education for English learners: Research-based approaches*. Sacramento, CA: California Department of Education.

Echevarria, J., Vogt, M.E., & Short, D. J. (2008a). *Making content comprehensible for English learners: The SIOP® Model* (3rd ed.). Boston, MA: Allyn & Bacon.

Echevarria, J., Short, D., & Vogt, M.E. (2008b). *Implementing the SIOP® model through effective professional development and coaching.* Boston, MA: Pearson/Allyn & Bacon.

Echevarria, J., Vogt, M.E., & Short, D. J. (2010). *The SIOP® Model for teaching mathematics to English learners*. Boston, MA: Allyn & Bacon.

Fathman, A. & Crowther, D. (Eds.). (2008). *Science for English language learners*. Arlington, VA: National Science Teachers Association.

Fellows, N. J. (1994). A window into thinking: Using student writing to understand conceptual change in science learning. *Journal of Research in Science Teaching, 31,* 985–1001.

Fisher, D., & Frey, N. (2008). *Word wise & content rich: Five essential steps to teaching academic vocabulary.* Portsmouth, NH: Heinemann.

Flynt, E. S., & Brozo, W. G. (2008). Developing academic language: Got words? *The Reading Teacher, 61*(6), 500–502.

Gandara, P., Maxwell-Jolly, J., & Driscoll, A. (2005). *Listening to teachers of English language learners: A survey of California teachers' challenges, experiences, and professional development needs.* Santa Cruz, CA: The Center for the Future of Teaching and Learning.

Garcia, G., & Beltran, D. (2003). Revisioning the blueprint: Building for the academic success of English learners. In G. Garcia (Ed.), *English learners: Reaching the highest levels of English literacy*. Newark, DE: International Reading Association.

Garcia, G. E., & Godina, H. (2004). Addressing the literacy needs of adolescent English language learners. In T. Jetton & J. Dole (Eds.), *Adolescent literacy: Research and practice* (pp. 304–320). New York, NY: The Guildford Press.

Genesee, F., Lindholm-Leary, K., Saunders, W., & Christian, D. (2006). *Educating English language learners: A synthesis of research evidence.* New York, NY: Cambridge University Press.

Gersten, R., Baker, S. K., Shanahan, T., Linan-Thompson, S., Collins, P., & Scarcella, R. (2007). *Effective literacy and English language instruction for English learners in the elementary grades: A Practice Guide* (NCEE 2007-4011). Washington, DC: National Center for Education Evaluation and Regional Assistance, Institute of Education Sciences, U.S. Department of Education. Retrieved from http://ies.ed.gov/ncee.

Gibbons, P. (2003). Mediating language learning: Teacher interactions with ESL students in a content-based classroom. *TESOL Quarterly, 37*(2), 247–273.

Goldenberg, C. (2008). Teaching English language learners: What the research does—and does not—say. *The American Educator, 32*(2), 8–23.

Gottlieb, M., & Lederman, N. (2006). Standards for science and English language proficiency. In A. Fathman & D. Crowther (Eds.), *Science for English language learners* (pp. 179–197). Arlington, VA: National Science Teachers Association.

Graham, S., & Perin, D. (2007). *Writing next: Effective strategies to improve writing of adolescents in middle and high schools.* A report to the Carnegie Corporation of New York. Washington, DC: Alliance for Excellent Education.

Hart, J., & Lee, O. (2003). Teacher professional development to improve science and literacy achievement of English language learners. *Bilingual Research Journal, 27*(3), 475–501.

Hiebert, E. (2008, April 30). *Critical science vocabulary: Challenges and assets.* Webinar and powerpoint. Available at http://www.schoolsmovingup.net/cs/smu/view/e/2628

Hiebert, E. H. (2005). *Word Zones™: 5,586 most frequent words in written English.* Available at www.textproject.org.

Hiebert, E. H. (2005). *1,000 most frequent words in middle-grades and high school texts.* Available at www.textproject.org.

Hiebert, E. H., & Lubliner, S. (2008). The nature, learning, and instruction of general academic vocabulary. In S. J. Samuels & A. Farstrup (Eds.), *What research has to say about vocabulary* (pp. 106–129). Newark, DE: International Reading Association.

Himmel, J., Short, D., Richards, C., & Echevarria, J. (2009). *Using the SIOP® model to improve middle school science instruction* (CREATE Brief). Washington, DC: Center for Research on the Educational Achievement and Teaching of English Language Learners. Available at www.cal.org/create/resources/pubs/siopscience.html.

Holliday, W. G., Yore, L. D., & Alvermann, D. E. (1994). The reading-science learning-writing connection: Breakthroughs, barriers, and promises. *Journal of Research in Science Teaching, 31*(9), 877–893.

Kagan, S. (1994). *Cooperative learning.* San Clemente, CA: Kagan Publishing.

Kindler, A. (2002). *Survey of the states' limited English proficient students and available educational programs and services. 2000-01 summary report.* Washington, DC: National Clearinghouse for English Language Acquisition.

Kober, N., Zabala, D., Chudowsky, N., Chudowsky, V., Gayler, K., & McMurrer, J. (2006). *State high school exit exams: A challenging year.* Washington, DC: Center on Education Policy.

Krashen, S. (1985). *The input hypothesis: Issues and implications.* London: Longman.

Lee, O. (2005). Science education with English language learners: Synthesis and research agenda. *Review of Educational Research, 75*(4), 491–530.

Lee, O. (2002). Science inquiry for elementary students from diverse backgrounds. In W. G. Secada (Ed.), *Review of research in education, Vol. 26* (pp. 23–69). Washington, DC: American Educational Research Association.

Lee, O., Deaktor, R. A., Hart, J. E., Cuevas, P., & Enders, C. (2005). An instructional intervention's impact on the science and literacy achievement of culturally and linguistically diverse elementary students. *Journal of Research in Science Teaching, 42*(8), 857–887.

Lee, J., Grigg, W., & Dion, P. (2007). *The nation's report card: Mathematics 2007.* (NCES 2007-494). US Department of Education, Institute of Education Sciences, National Center for Education Statistics. Washington, DC: U.S. Government Printing Office.

Lee, J., Grigg, W., & Donahue, P. (2007). *The nation's report card: Reading 2007.* (NCES 2007-496). US Department of Education, Institute of Education Sciences, National Center for Education Statistics. Washington, DC: U.S. Government Printing Office.

Lemke, J. (1990). *Talking science: Language, learning and values.* New York, NY: Ablex.

Marzano, R. J., & Pickering, D. J. (2005). *Building academic vocabulary for student achievement: Teacher's manual.* Alexandria, VA: Association for Supervision and Curriculum Development.

Mehan, H. (1979). *Learning lessons*. Cambridge, MA: Harvard University Press.

Meredith, K., Meredith, J., & Temple, C. (1997). The insert method. In *Reading and Writing for Critical Thinking (RWCT) workbook.* Newark, DE: International Reading Association.

Nagy, W., & Scott, J. (2000). Vocabulary processes. In M. Kamil, P. Mosenthal, P. D. Pearson, & R. Barr (Eds.), *Handbook of reading research*, Volume III (pp. 269–284). Mahwah, NJ: Erlbaum.

National Center for Education Statistics. (2002). Schools and staffing survey, 1999-2000: *Overview of the data for public, private, public charter, and Bureau of Indian Affairs elementary and secondary schools.* (NCES 2002-313). Washington, DC: U.S. Department of Education, National Center for Education Statistics.

National Institute of Child Health and Human Development (NICHD). (2000). *Report of the National Reading Panel, Teaching children to read: An evidence-based assessment of the scientific research literature on reading and its implications for reading instruction.* (NIH Publication No. 00-4769). Washington, DC: U.S. Department of Health and Human Services.

Oczkus, L. (2009). *Interactive think-aloud lessons: 25 surefire ways to engage students and improve comprehension*. New York: Scholastic, and Newark, DE: International Reading Association.

Parish, T., Merikel, A., Perez, M., Linquanti, R., Socias, M., Spain, M., Speroni, C., Esra, P., Brock, L., & Delancey, D. (2006). *Effects of the implementation of Proposition 227 on the education of English learners, K–12: Findings from a five-year evaluation.* Palo Alto, CA: American Institutes for Research.

Reiss, J. (2008). *102 content strategies for English language learners*. Upper Saddle River, NJ: Pearson/Merrill Prentice Hall.

Rosebery, A., & Warren, B. (Eds.). (2008). *Teaching science to English language learners.* Arlington, VA: National Science Teachers Association Press.

Rowe, M. B. (1996, September). Science, silence, and sanctions. *Science and Children, 34*(1), 34–37.

Saul, E. W. (2004). *Crossing borders in literacy and science instruction.* Newark, DE: International Reading Association/National Science Teachers Association.

Saunders, W., & Goldenberg, C. (in press). Research to guide English language development. In D. Dolson & L. Burnham-Massey (Eds.), *Improving education for English learners: Research-based approaches.* Sacramento, CA: California Department of Education.

Scott, J. (Ed.). (1992). *Science and language links: Classroom implications.* Portsmouth, NH: Heinemann.

Scott, J. A., Jamison-Noel, D., & Asselin, M. (2003). Vocabulary instruction throughout the day in twenty-three Canadian upper-elementary classrooms. *The Elementary School Journal 103*, 269–286.

Short, D. (2009). *Sheltered instruction: Curriculum and lesson design.* Paper presented at the 31st Sanibel Leadership Conference, Sanibel, FL, June, 2009.

Short, D., & Fitzsimmons, S. (2007). *Double the work: Challenges and solutions to acquiring language and academic literacy for adolescent English language learners.* Report to Carnegie Corporation of New York. Washington, DC: Alliance for Excellent Education.

Short, D., & Hillyard, L. (2005). *SIOP® unit planner.* Unpublished manuscript. Washington, DC: Center for Applied Linguistics.

Short, D., & Richards, C. (2008). *Linking science and academic English: Teacher development and student achievement.* Paper presented at the Center for Research on the Educational Achievement and Teaching of English Language Learners (CREATE) Conference, Minneapolis, MN, October, 2008.

Short, D., & Thier, M. (2006). Perspectives on teaching and integrating English as a second language and science. In A. Fathman & D. Crowther (Eds.), *Science for English language learners* (pp. 199–215). Arlington, VA: National Science Teachers Association.

Short, D., Fidelman, C., & Louguit, M. (2009). *The effects of SIOP® Model instruction on the academic language development of English language learners.* Manuscript submitted for publication.

Short, D. J., Vogt, M.E., & Echevarria, J. (in press). *The SIOP® Model for teaching history-social studies to English learners.* Boston, MA: Allyn & Bacon.

Siegel, H. (2002). Multiculturalism, universalism, and science education: In search of common ground. *Science Education, 86*, 803–820.

Snow, C.E., Cancino, H., De Temple, J., & Schley, S. (1991). Giving formal definitions: A linguistic or metalinguistic skill? In E. Bialystok (Ed.), *Language processing and language awareness by bilingual children* (pp. 90–112). Cambridge: Cambridge University Press.

Stahl, S. A., & Nagy, W. E. (2006). *Teaching word meanings.* Mahwah, NJ: Erlbaum.

Suarez-Orozco, C., Suarez-Orozco, M. M., & Todorova, I. (2008). *Learning in a new land: Immigrant students in American society.* Cambridge, MA: Harvard University Press.

Tharp, R., & Gallimore, R. (1988). *Rousing minds to life: Teaching, learning and schooling in social context.* Cambridge, MA: Cambridge University Press.

U.S. Department of Education. (2006). *Building partnerships to help English language learners.* Fact sheet. Retrieved January 2, 2008 at http://www.ed.gov/nclb/methods/english/lepfactsheet.html.

Vogt, M.E., & Echevarria, J. (2008). *99 ideas and activities for teaching English learners with the SIOP® Model.* Boston, MA: Allyn & Bacon.

Walqui, A. (2006). Scaffolding instruction for English language learners: A conceptual framework. *The International Journal of Bilingual Education and Bilingualism, 9*(2), 159–180.

Watson, K., & Young, B. (1986). Discourse for learning in the classroom. *Language Arts, 63*(2), 126–133.

Zehler, A. M., Fleishman, H. L., Hopstock, P. J., Stephenson, T. G., Pendzik, M. L., & Sapru, S. (2003). *Descriptive study of services to LEP students and to LEP students with disabilities; Policy report: Summary of findings related to LEP and SpEd-LEP students.* Arlington, VA: Development Associates.

Zwiers, J. (2004). *Developing academic thinking skills in grades 6–12.* Newark, DE: International Reading Association.

Zwiers, J. (2008). *Building academic language: Essential practices for content classrooms.* San Francisco, CA: Jossey-Bass.

index